THE DIVINE UNIVERSE

Other books by Swami Abhayananda:

The Supreme Self
History of Mysticism: The Unchanging Testament
Jnaneshvar: The Life And Works
Dattatreya: Song of The Avadhut
Thomas á Kempis: On The Love of God
The Wisdom of Vedanta
Plotinus: The Origin of Western Mysticism
Mysticism And Science: A Call for Reconciliation

THE DIVINE UNIVERSE

An Alternative to The Scientific Worldview

by

Swami Abhayananda

iUniverse, Inc.
New York   Bloomington

The Divine Universe
An Alternative To The Scientific Worldview

Copyright © 2008 by Swami Abhayananda

All rights reserved. No part of this book may be used or reproduced by any means, graphic, electronic, or mechanical, including photocopying, recording, taping or by any information storage retrieval system without the written permission of the publisher except in the case of brief quotations embodied in critical articles and reviews.

The views expressed in this work are solely those of the author and do not necessarily reflect the views of the publisher, and the publisher hereby disclaims any responsibility for them. This book may not be reproduced in whole or in part without prior written consent from the author.

iUniverse books may be ordered through booksellers or by contacting:

iUniverse
1663 Liberty Drive
Bloomington, IN 47403
www.iuniverse.com
1-800-Authors (1-800-288-4677)

Because of the dynamic nature of the Internet, any Web addresses or links contained in this book may have changed since publication and may no longer be valid. The views expressed in this work are solely those of the author and do not necessarily reflect the views of the publisher, and the publisher hereby disclaims any responsibility for them.

ISBN: 978-0-595-52751-9 (pbk)
ISBN: 978-0-595-62985-5(cloth)
ISBN: 978-0-595-62803-2 (ebk)

Printed in the United States of America

Table of Contents

INTRODUCTION — 1
I. CONSCIOUSNESS AND ENERGY — 5
1. MYSTICISM, SCIENCE, AND THE HEIRS OF DEMOCRITUS — 7
2. THE ULTIMATE THEORY OF EVERYTHING — 17
3. THE ORIGIN OF THE UNIVERSE — 20
4. WHAT IS ENERGY? — 26
5. WHERE CONSCIOUSNESS COMES FROM — 32
6. TIME, ETERNITY, AND THE FUTURE TASK OF SCIENCE — 37
7. IN THE FINAL ANALYSIS — 41

II. SPIRITUAL VISION — 45
8. AGNOSTICISM EXAMINED — 48
9. HOW DO WE KNOW? — 51
10. ENLIGHTENMENT AND GRACE — 55
11. MY OWN EXPERIENCE — 63
12. THE GIFT OF SPIRITUAL VISION — 74
13. WE WHO HAVE BEEN BLESSED — 77
14. HE HEARS — 78

III. THE PERENNIAL PHILOSOPHY — 81
15. KAPILA'S VISION — 83
16. THE PHILOSOPHY OF NONDUALISM — 88
17. PERFECT NONDUALISM — 93
18. OX-HERDING — 114
19. THE APPEARANCE OF DUALITY — 123
20. NONDUALISM IN THE TEACHINGS OF JESUS — 127
21. THE MEETING OF HEART AND MIND — 142

IV. THE SCIENCE OF THE SOUL — 149
22. THE ASTROLOGY OF ENLIGHTENMENT — 151
23. THE RATIONALE FOR ASTROLOGY — 164
24. THE SOUL OF ASTROLOGY — 169
25. ASTROLOGY AND FREE WILL — 175
26. ETERNAL FREEDOM — 185

27. APPENDIX 1: 187
REFLECTIONS ON THE TWO DEFINITIONS OF ENERGY.. 187

28. APPENDIX 2: 193
WHAT IS A SWAMI? 193
A SONG OF THANKSGIVING............................. 197
About the Author 199

Dear Father, Lord of the universe, Guide and Protector of all Thy children, Thou knowest that this body, heart, mind and soul are Thine own; please do whatsoever Thou wilt with them. And if it please Thee, let my words be to Thy honor and to Thy glory! And may they benefit all Thy children. I whisper this prayer close to Thy ear in hopes Thou wilt grant it. And since Thou art the only 'I', it seems clear that Thou art speaking to Thyself in this wish. Please grant the strength and wisdom to this body, heart, mind and soul that is required to carry out this wish of Thine. So may it be.

INTRODUCTION

For many, contemporary materialistic science offers a sufficiently convincing worldview; but I wish in this book to offer an equally convincing alternative to that established worldview. I offer not a refutation, but rather a reformulation of the scientific perspective into which the worldview of spirituality is neatly integrated. Upon examination, this worldview integrating science and spirituality will be recognized to be an ancient and perennial one, and yet it is a vision that wonderfully satisfies the requirements and sensibilities of the modern intellect as well. However, it is a vision that can only be approximated, and never *fully* told. For, to be truly known, its truth must be revealed to the inner eye, and thus "seen" by each single soul who seeks to know it. It is a vision that does not lend itself well to language, but shines forth and communicates itself clearly through a higher and subtler means of expresssion that is at once intuitive and revelatory. And so these words I offer in the service of Spirit are only suggestive, like the finger pointing at the moon. Only the reader can make them productive of understanding by tracing their meaning to the living Reality within to which they point.

In our contemporary world, the spiritual worldview is very much under attack. Many books have appeared on the market today touting scientism and decrying the spiritual worldview, and just the other day, I heard a segment on the radio highlighting a group of atheists. How smug they seemed with their scientific perspective on things, and how condescending they were toward those they referred to as "believers", we poor ignorant masses of superstitious humanity. I could only laugh. Years ago, as a young man, I sympathized with their position. I saw no evidence for belief in God; in fact, those who embraced religion seemed to me to be merely passive followers of the naïve beliefs blindly accepted by the culture as a whole. When I was twenty-eight, however, my mind became opened to the possibility of the direct experience of God, and I went into solitary retreat in a mountain cabin to prepare myself for a direct meeting with God. By the grace of God, that meeting came on the night of November 18, 1966.

At that time, drawn deeply into contemplative prayer, I experienced from the vantage of eternity the outflow of the universal manifestation and its subsequent return in a never-ending cycle of manifestation and dissolution. Much later, I read of the theory of 'the Big Bang' put forward by the theoretical physicists. It was not long before I realized that the initial expansion of the newborn universe, said by the physicists to have occurred around 15 billion years ago from an 'infinitely dense point', was the same origin that I had witnessed in meditation years earlier. With this understanding, I set out to reconcile these two visions—one from the viewpoint of the Eternal, and one from the viewpoint of contemporary theoretical physics—in the hope of bringing about a synthesis of the spiritual and the scientific visions regarding the origin of our Cosmos.

Here, then, is a collection of independent Essays on various aspects of this integrated worldview, written spontaneously over the past year or so, with an intent to offer a clear and reasoned alternative to the worldview promulgated by the many advocates for the popular 'scientism' of our age. There are four distinct 'groups' of Essays included here: there are those that deal with correcting some of the myths of popular science; there are some that are expressive of the 'perennial philosophy'; there are some that deal with that much maligned subject: astrology; and there are those which attempt to give some idea of what it is like to "see God" (See Chapter 11, "My Own Experience").

One of the reasons for the difficulty in describing such an experience is the fact that God is not experienced as someone or something that can be spoken of in the third person as "He" or "Him", or even spoken of in the second person as "Thou" or "Thee". God is experienced as one's Self, and therefore can only be spoken of as "I". In the religious traditions of India, this understanding is commonplace; God is spoken of as *Paramatman*, "the Supreme Self", or simply as the congregation of the subjective qualities *sat*, "Being or Existence"; *chit*, "Consciousness"; and *ananda*, "Bliss". Yet in our Western culture and language, this entanglement of the individual's "I" (or *ego*) and the Divine "I" still makes for confusing and problematic communication regarding the subject of God, the Divine Self.

Perhaps the most persistent and perplexing question about God is "How is the experience of God to be attained? Is there a reliable scientific answer to the question of how this can be done?" And the answer is "No". To be sure, the focused directing of the soul's attention to the eternal Reality through meditation or prayerful contemplation is paramount; but why do so few obtain the desired results where so many make the effort? There are clearly no clear cut guidelines that can promise success in this endeavor. And so it has always been regarded as a matter of God's grace or favor. This declaration of partiality on the part of God is regarded by many as unsatisfactory, though individual merit does not seem to be a determining factor either. Yet, how else may we regard it? It is possible that the karmic evolution of the soul is a factor. Having discovered some unusual planetary phenomena occurring at the time of my "mystical" experience, I have suggested the possibility of a connection between the two occurrences; but the establishment of a tangible correlation between them awaits the collection of data concerning many more such experiences. The fact is that we do not know for sure why God reveals Himself in some and not in others.

The question of how a God who is eternal Consciousness is able to "create" this immense and multi-faceted universe is also one which presents a stumbling block for many. From my own experience, the universe is projected and withdrawn in a recurring cycle, in the manner of a breath that is exhaled and inhaled. Each cycle of that 'breath' lasts, from our temporal perspective, for billions of years; yet from the perspective of eternity, beyond time and space, each lasts for merely the space of a breath. God is not confined to human possibilities; He is at once eternally transcendent Consciousness, and active Energy operating in the spatio-temporal field. He is both unmoved and mover. He projects or emanates our universe in a manner similar to the way we project a thought-form or dream upon our own consciousness while remaining the witness to our creations.

Underlying a dream phantasm is the active mind of the dreamer. That dreamer's mind is the material cause, the formal cause, the effective cause and the final cause of the dream. Using that analogy, God, the Divine Mind whose projected "dream" this universe is, is the material, formal, effective and final cause of this phenomenal world. Once this is grasped, what further purpose does the investigative analysis of this world serve? It brings to mind the thought of a scientist-character in a dream tearing up the dream-pavement in the dream-landscape in order to analyze it, then placing the pieces under a dream-microscope. We might further imagine such a dream-scientist coming up with pronouncements about what this dream-terrain is made of, such as: "It seems to be made of waves!" "No, it is made of particles, but the particles themselves seem to be nothing more than a kind of energy!" "I'll be damned! It's both waves and particles! What *is* this stuff?" Truly, it is clear that such efforts would be utterly futile, and that, in order to really know the truth about himself and the reality in which he lived, our dream-scientist would simply need to wake up. Our dreams thus show a close parallel to the nature of our 'real' universe. While I do not wish to denigrate the efforts of scientists, I have seen that the true nature of 'reality' can only be realized by those who 'wake up' to the eternal Self.

While that eternal Self is forever unaffected by the evolution of our cosmos, He is intimately involved in it. Just as our own consciousness is involved in the play of dreams, so is the one Divine Consciousness playing in this universal drama. He is the Self of our self, the Joy of our joy; and as we evolve toward full awareness of His truth, our understanding will eventually become clearer and expand to encompass both the heavens and the earth. I sincerely hope that the following collection of Essays will stimulate you to look deeply into the nature of your own self and the universe around you, and truly come to see yourself as the one Divine Consciousness playing in your own Divine Universe.

☙ ☙ ☙

I.

CONSCIOUSNESS AND ENERGY

Let me say at the onset that I have no scientific training. My interest in cosmogony derives primarily from my own direct "mystical" experience. I certainly would not pretend to know anything about this universal 'Creation' if I had not seen it in the light of an inner revelation, while drawn into a deep contemplative union with the Father. That said, I have also immersed myself deeply and for a long period of time in the current vision of science in order to comprehend that perspective as well as I am able. And I am now attempting to bring together my vision of gnosis with the vision of science in the hope of shedding some small amount of revealing light on both.

The theology of the illumined mystics is the same the world over. Only the names for God and His Power are different owing to the differing languages. All hold that the Supreme Being is absolute and unchanging. And all hold that He possesses a creative Power by which He manifests this spatio-temporal universe. In His eternally absolute and unchanging aspect, He has been called by one name, and in His aspect of universe Creator, He is called by another name. In the West, these two apsects of God have been called Theos *and* Logos, Jahveh *and* Chokmah, The One *and* Nous, Godhead *and* God, Father *and* Mother, *and so on. In the East, they have been called* Prajapati *and* Prthivi, Purusha *and* Prakrti, Shiva *and* Shakti, Brahman *and* Maya, Tao *and* Teh, Haqq *and* Khalq, *and many other names. In our modern era, the names most commonly used to denote these two aspects of God are the* Divine Consciousness *and the* Divine Energy.

Undoubtedly, some confusion arises due to the fact that these terms, consciousness *and* energy, *are also used by contemporary scientists in a more limited context to denote quite different realities. For example, science does not recognize Consciousness as the universal Source of all, but rather sees it as a mysterious byproduct of the biological activity of the human brain. Likewise, the term, Energy, which I use in its theological sense as the Divine Power, has a traditional use in the scientific lexicon as an ambiguously defined term attached to various qualifiers—chemical, nuclear, thermal, potential, electrical, etc.—to represent the dynamic activities of these differing material frameworks. And so there is a paradigmatic disconnect between the conceptions and terminology of theology and science, as they are quite different both in content and meaning. And so, here, in this First Section, I present some thought-provoking Essays regarding* the contemporary scientific perspective, and some innovative ideas on how this perspective might be enhanced by the perspective of gnosis.

1.
MYSTICISM, SCIENCE, AND THE HEIRS OF DEMOCRITUS

Part One

Mysticism and science represent two opposing worldviews which may be reduced to the two diametrically opposed philosophical positions known as *idealism* and *materialism*. These two starkly differing views of the nature of the reality underlying the appearance of the world have been at odds with each other for twenty-five centuries beginning with Pythagoras, Xenophanes, Anaxagoras and Socrates on the idealist side, and Thales, Leucippus, and Democritus on the materialist side. Idealists hold that Mind is the primary reality of which matter is an evolute; materialists hold that matter is the primary reality of which mind is an evolute. Mystics, those who claim to have actually experienced or "seen" the ultimate reality directly in a moment of contemplative revelation, fall squarely on the side of idealism. Every mystic who ever lived has declared the idealistic viewpoint, stating that the ultimate reality underlying all phenomena is unquestionably noumenal; i.e., a transcendent Mind. There are no materialists among mystics.

Mysticism, therefore, is an idealist point of view which asserts the possibility of the *direct* apperception of the ultimate reality in a rare, profound, and purely introspective experience, wherein an extraordinarily intimate knowledge of the noumenal Source and the nature of the universe and human existence is acquired. This "mystical experience", say those who have known it, reveals the formless, transcendent Noumenon, the "groundless Ground" of all physical and mental phenomena, which is seen to constitute everyone's original and eternal identity. Such an experience seems to have been first spoken of in ancient Greece among the populace taking part in the "mystery religions" such as the Eleusinian and Orphic mysteries (whence mysticism gets its name); and later formed the basis of the philosophical position of such seers as Socrates (by way of Plato), Philo Judaeus, and Plotinus. In the East, mysticism made its appearance in the writings of Lao Tze, the Upanishads, and the early Buddhist texts, and later in the Middle East with the teachings of Hermeticism, and the rise of Christianity and Gnosticism, all of whose central figures claimed an intimate, mystical knowledge of the noumenal Source.

Science, in its present state, represents the position of materialism; though, it should be noted, science is not *necessarily* materialistic; that is, materialism is not an *essential* feature of science, shown by the fact that many of the greatest scientists who ever lived held religious views which demanded a noumenal source for the phenomenal world. But there is an established trend among modern scientists toward an exclusively materialistic view, no doubt as a result of the emphasis in science on conclusions which are empirically demonstrable. Science deals in tangibly objective sense data, and does not comfortably extend to less tangible subjective mental states. The very definition of science limits its focus to only that which may be empircally verified. And that requirement assures that science will probably always tend to have a materialistic bias, and will grant little credence to noumena experienced in a subjective and unverifiable state of awareness.

While science, and its attendant materialism, may be said to have originated with the early Greek philosophers cited above, it had to struggle in the West for many centuries against the strictures of religious doctrine, and only began its cultural ascendency from the seventeenth century onward, influenced by such philosophers as Francis Bacon, Thomas Hobbes, John Locke, David Hume, and Immanuel Kant, and the works and accomplishments of scientists such as Galileo, Johannes Kepler, and Isaac Newton. By the twentieth century, materialism was firmly embedded in the scientific (empirical) method and implicitly formulated in the widely held philosophy of logical positivism. This view, that only knowledge obtained by the scientific method and capable of being demonstrated experimentally was worthy of the label 'knowledge', became the widespread faith of our Western culture, a faith referred to by its critics as 'scientism'. And, while there are still a few maverick idealists among the ranks of scientists today, the vocal majority utterly reject the slightest hint of mysticism or idealism, and hold as firm doctrine that the universe came into being and is sustained through "natural," that is to say, purely material, processes. Nevermind that "matter", upon close examination, dissolves into "thought".

These two, empirical knowledge, or *science*, and mystical knowledge, or *gnosis*, represent knowledge obtained through two radically different methodologies: empirical knowledge represents the ordering and analysis of *outward* observations of phenomena perceived by the senses in the normal waking state; mystical knowledge represents the *inward* observation of noumena intuitively perceived by the mind in a highly extraordinary, but well documented, contemplative state. They are really two different kinds of knowledge, referred to as *science* and *gnosis*. *Science* is from the Latin *scientia*, derived from *scire*, to know, and usually denotes the organization of objectively verifiable sense experience; *gnosis* is a Greek word, also meaning knowledge, but denoting an inwardly "revealed" knowledge unavailable to science.

The difficulty presently apparent is that advocates of materialistic science refuse to acknowledge not only the validity and relevance of gnosis, but even the very possibility of its existence. Today, science is so steeped in the materialistic perspective that scientists and, through their influence, "educated" members of the public, routinely regard all those who hold to idealistic views as unfortunate members of the ignorant and uneducated masses, misguided by superstition. Those with a mystic bent are held in especial disdain, and are the subjects of frequent ridicule in our materialist oriented culture. Colleges and universities around the nation instill this arrogant prejudice in the youth who flock to them for their one-sided educations. One has to wonder if we are not due at this time in our history for a return of the cultural pendulum to a fresh idealism, one that is informed by both science *and* gnosis.

Part Two

Let's go back once again and look a little closer at the initial split between these two ways of knowing: It probably began with the earliest hominids; but the best records of this division that we possess from Western civilization only go back around twenty-five hundred years to ancient Greece. Democritus (ca. 460-390 B.C.E.), student of Leucippus, contemporary of Socrates, was the Greek philosopher who surmised that the world we live in is made up of very small, indivisible, entities which he called *Atoms*. These atoms, he guessed, were the elementary particles and building blocks of the cosmos, and were, therefore, the ultimate and final answer to the question 'what is everything made of?' Democritus was a firm materialist. He was, in fact, the foremost in a long line of 'materialistic scientists'. He saw no need to look any further than these 'elemental' particles for the material foundation of existence. Other materialists of the time were Thales (ca. 625-545 B.C.E.), who thought that water was the 'material principle' of the world; and Anaxamenes (fl. 548 B.C.E.), who believed that the element, air, was the fundamental constituent of everything. But there were some other philosophers of the period who were a bit more intuitional, and certainly more contemplative, in their approach to the knowledge of ultimate reality. These philosophers had "seen" into the depths of their own conscious minds, and discovered through that vision that the source of the material universe is not itself material, but is rather an eternal Mind, a Noumenon beyond all phenomena, who is the source of the phenomenal, projecting the cosmos as a human mind projects thoughts and ideas upon itself. This view, known as

The Divine Universe

idealism, was held by Xenophanes (ca. 580-480 B.C.E.), Pythagoras (b. 570 B.C.E.), Parmenides (b. ca. 540 B.C.E.), Anaxamander (fl. 547 B.C.E.),), Heraclitus (fl. ca. 500 B.C.E.), and of course Socrates (469- 399 B.C.E.) and Plato (427-347 B.C.E.).

To Those more contemplative and introspective philosophers, a reality underlying the material appearance was revealed. That reality, they said, is an eternal Mind, or Consciousness, which produces all of the phenomenal (material) universe by Its manifestory power (Divine Energy). That Mind was revealed to those contemplative philosophers as the Ground of all; but the materialists could not 'see' that Ground; they saw only the figure. To the introspective philosophers, the eternal Mind, the Father, was revealed as the sole Reality in whom and by whose creative Power this universe exists. That eternal Mind, they said, cannot be seen by the eyes; but It makes Its Energy visible in the form of matter. However, for the materialists, Energy (matter) is all that exists.

It seems that, after 2500 years, the controversy is still unresolvable. Both the materialistic scientist, Democritus, and the idealists such as Socrates and Plato, have their present-day descendents who continut to battle in this conflict. Some consider the reason for this division in human perspectives to lie in the differences in the educations and life-experiences—in other words, the nurture—of those individuals making up these two philosophical worldviews. Others feel that it may be because of certain basic differences in the cerebral makeup—in other words, the nature—of idealists and materialists. Perhaps there are subtle differences related to the evolutionary stage at which each individual soul finds itself; perhaps these differences are reflected in right-brain/left-brain patterns of dominance. Perhaps it is simply a matter of Grace. Who can say? But what is certain is that this duality of philosophical perspectives greatly affects our current society and colors nearly every aspect of the conduct of life on earth.

In our contemporary American culture, these opposing views may exist unnoticed side by side, often within the same individual. Many find that their favorite religious faith provides their subconscious idealistic perspective, while their worldly preoccupations bespeak their conscious materialistic bias. But these two co-existing, though opposing, ideologies are rarely ever analyzed, defined or even mentioned in our society. Religious faith and materialistic science co-exist comfortably within the minds of the vast majority of the indiscriminant masses. In fact, materialistic science, and its corollary, 'scientism', has for several centuries been sanctified as the ideology of choice within the American culture. And though we, as a culture, currently seem to be slowly emerging from that lengthy period of blind materialism, the materialistic perspective continues to flourish, and no doubt shall continue until the last man and child on earth becomes enlightened by the merciful grace of God.

Today, there are many heirs to Democritus' materialistic science who are vociferous in extolling their ideology. I would like to mention two of them, without mentioning their names: One is a theoretical physicist, physics professor, and best-selling author. In his latest book he attempts to enthuse his reading audience for the expected coming validation of 'Superstring Theory', which, he expects, will prove that the ultimate reality is actually very tiny material 'strings' of which all matter and forces are made. It seems that someone has calculated mathematically that the present menagerie of particles and forces so far discovered may be reduced to a common unifying 'element' if all those particles and forces were themselves constituted of a yet tinier material entity in the form of vibrating strings, which would then, according to theorists, produce by their vibrations and varying configurations the appearance of every particle and force thus far known. The only problem is that these 'strings' would have to be so tiny that, if a hydrogen atom were blown up to the proportions of the Milky Way galaxy, strings within it would only be the size of dust mites. It would take more than a billion, billion quadrillion of these strings to make up an inch. Also, they would have to exist in a universe consisting of ten to twenty-four curled-up dimensions.

Wouldn't it be wonderful if you really could infer the ultimate reality by taking things apart and finding that one common element in everything! However, it's a very multi-faceted and insubstantial ocean of constantly transforming (Thought) energy that we find instead. The cosmos in which we live almost seems to be designed in such a way as to confound any and all efforts to comprehend the manner of its existence. Fortunately, the One who is the ultimate Source of this energetic ocean of appearance has periodically revealed Himself to certain individuals and made known the manner of His projection of this universal array. But, unfortunately, that vision and that certainty is not available to all. There's the rub. So the unillumined go on refusing to acknowledge a Mind greater than their own; and they go on inventing myriads of incredibly bizarre scenarios for the origin and constituency of the universe. They go on enquiring, delving, analyzing, and presupposing, wending their way more and more deeply and inextricably into labyrinthine mazes of imagination – all to no avail. Isn't it amazing what an ingeniously designed comedic drama the Author of this universal production has fostered![1]

Another materialistic scientist, a Cosmologist, also a professor and author, is anxiously awaiting the empirical verification of the 'quantum fluctuations in the vacuum of space' as the ultimate cause and origin of the 'Big Bang'. He suggests that the universe began from nothing as a "quantum fluctuation in the vacuum"; but it seems to me that one would then be required to explain what caused the quantum vacuum. Is the "quantum fluctuation" the prime mover, the ultimate reality? I'm being facetious, of course; I know it's not the ultimate reality. I've seen the ultimate Source. He lives in/as eternity, and this universe is the projection of His will, an indescribable breathing forth of the whole Mind-born shebang and a subsequent withdrawing of it all once again, a cycle endlessly repeated. Why? No one knows. And I don't think there is a why comprehensible to us. But the important point is that, while the manifested universe is our temporal reality, that one Mind is our eternal reality. And He can be known within as the consciousness of "I" through His gracious revelation.

In a recent book, our Cosmologist offers ten questions which comprise his ten Chapter titles: *1. How do we know the things we think we know? 2. Is there a theory of everything? 3. How did the universe begin? 4. How did the early universe develop? 5. Why is the universe the way it is? 6. What is it that holds the universe together? 7. Where did the chemical elements come from? 8. Where did the solar system come from? 9. Where did life originate? 10. How will it all end?* While our Cosmologist explains the answers to each of these questions as 'natural' processes, I couldn't help laughing when I realized that, for me, in my simplistic view of things, the answer to each of these questions is perfectly obvious. The answer to each is "God". Needless to say, that answer would fall short of satisfying any of our materialistic scientists. But it clearly points out the immense difference between our perspectives on reality.

For me, the richness of the multitude of universal phenomena is understood to be projected by, and contained within, the One. The One, and not the perplexing multitude of phenomena, is the unvarying focus of my attention. Having seen the splaying out of the universe from the vantage of eternity, curiosity for just how each particular phenomenon is produced is utterly lacking in me. What a simple bumkin I must seem! Yet I truly believe that, once the scientists follow all their theoretical extrapolations to their ultimate resolution, they will come at last to the same simple unity in which I am comfortably settled. They may call it by another name, but they must in the end come to the one eternal Mind that has breathed forth this immensely complex universe of seething motion. That is the ultimate Theory of Everything. The universe began from (in) Him. The universe is the way it is because He thought (willed) it so. It is His Thought that produced it and holds it together. The chemical elements, the solar system, and life all come from Him. It will end also by His will when He withdraws it all back into Himself.[2] This is the theory backed up by the visionary experience of countless mystics, seers, sages, and prophets from time immemorial.

In the conceptualization of a materialistic universe produced by a material cause, there are clearly no limits to the possibilities of one's imagination. These clever materialistic scientists hope one day to announce to the world: 'We've finally discovered what the universe is made of; it's made of a whole lot of strings!' 'And it all began with a random fluctuation!' But, sorry boys; you're on the wrong track. We (mystics) have seen the ultimate source, and turns out He's an eternal Mind, who, though completely beyond our time and space universe, also intimately pervades and constitutes this universe as divine Thought. That's why you keep coming up with little particles that turn out to be waves of pure (Thought) energy. That's why all those little particles seem to be interconnected, though there is nothing apparently connecting them. That's why you can't get a handle on what's making the whole thing hold together and behave as an intelligently guided and integral whole. That's why you're never going to discover the ultimate reality by means of a microscope or telescope or supercollider. Give it up, boys. The ultimate reality is an open secret already; and you guys have been sadly and terribly misled by your unillumined mentors. It's okay if you're just clowning around, trying to see what amazing fantasies you can come up with; go ahead, knock yourselves out. But please give some due acknowledgment and respect to the truth as it has already been revealed countless times to countless individuals.

The two materialistic scientists cited above are certainly typical and representative of the present viewpoint of many scientists both in America and abroad. But we mustn't imagine that there are no exceptional scientists who reject the materialistic bias of their many associates. One such exceptional scientist is a professor emeritus of the University of Minnesota at Minneapolis by the name of Roger S. Jones. His latest book, *Physics For The Rest of Us*,[3] raises many pertinent questions regarding the phenomenon of the idolatry of science evident in the West, and has much advice to offer on the side of caution. He asks, "Can we afford to maintain the separation of science from philosophy and religion? Can we continue to judge science apart from its ethical and aesthetic implications? Can we tolerate a science that categorically denies human meaning, value, and purpose?"[4]

And he gives an impressive answer to these questions:

"THE LIGHT OF THE LAMPPOST

Despite the long-standing and pervasive practice in the West, there is nothing natural or essential about separating the humanities from the sciences. We have already explored the common origin that science and religion shared in the human quest to find the meaning and purpose of existence. In earlier times, this search was treated holistically. What we think of separately as spiritual and physical matters were formerly considered one unified area of knowledge. In mythology, for example, divine influences and interventions commonly determine matters on the earthly plane. In Platonic philosophy, there is an essential bond between the ideal realm and the physical plane. Indeed, the study and contemplation of things physical is supposed to enlighten human beings and lead them to the spiritual realm of Platonic Ideas.

"With Aristotle, however, things began to change. Although there were still important connections between the divine celestial spheres and the sublunary realm, there was a growing emphasis on the knowledge of physical and biological phenomena on the earth. Why was this so?

"In the effort to make sense of existence, human beings sought order and meaning in their environment—in the stars, elements, plants, and animals. But although the original motivation was to find a rationale for human existence, a gradual shift of emphasis took place. The explanation of the ways of the gods to man remains a vague and problematic task. It is subject to individual interpretation, inspiration, and revelation. There are no final answers. It is a frustrating and taxing quest with no easy rewards.

"On the other hand, the study of the purely *material* aspects of natural phenomena, unencumbered by the effort to interpret their divine or spiritual meaning, is a less frustrating and less ambiguous task with more immediate rewards. Describing the chemistry and biology of a rose has turned out to be more definable, achievable, and practical than attempting to divine the cosmic plan behind the rose in the first place. And so Aristotle and many who came after him began to emphasize material studies over metaphysical matters.

"It's like the ironic tale of the man who searches for his lost keys under a lamppost, not because he lost them there but because there's more light there to see by. What we are capable of doing most efficiently and effectively will often sway us and make us forget the more difficult task that we set out to accomplish in the first place. Scientific work is not easy. But it has certain appealing, gratifying, and rewarding characteristics that are extremely rare or entirely lacking in such fields as theology and philosophy—a sense of immediacy and verifiability, a level of consistency and reproducibility, a feeling of contact with reality, a history of practical achievements, a well-defined mathematical language of description and prediction, an explicit methodology of procedures and techniques, an inherent intelligibility and communicability, an evolving and cumulative sense of progress, and involvement with the affairs of societies and nations, and an unprecedented aura of prestige and authority.

"But for all its brilliant traits, fabulous techniques, and shining achievements, science has not brought us one jot closer to fathoming the human condition and the mystery of existence. Science has greatly ameliorated life on earth and has given us vast power and control over nature. That no one can deny. But if it is our souls and the meaning of life we seek, then the light of the lamppost has illuminated nothing." [5]

Later, he adds: "The fabulous successes of physics in the nineteenth and twentieth centuries provided a dramatic contrast with the continuing frustrations of philosophy and the general decline of religion and theology. Science—and especially physics—became king and remains so to this day. The light of the lamppost has blinded us all." [6]

It has not blinded *all* of us. There are a few of us who are illumined, not by the lamppost, but by the eternal Light. It is by that Light that we are able to see the Truth of the universe and ourselves. In fact, I believe that we humans are now, at this moment, in the midst of a significant shift in our collective understanding. In the eighteenth and nineteenth centuries, it became clear that metaphysical speculations, however well reasoned, were untestable and unreliable; that the very attempt to discover the hidden spiritual reality by means of the reasoning mind was a fool's task. That established, empirical Science quickly rose to the ascendency, providing what many thought of as a foolproof means of determining the facts, as well as they could be discovered.

For many, we are still in that watershed stage of distinguishing empirical science from metaphysical speculation; but there is a new kid on the block, that slowly began to reveal its presence around the middle of the twentieth century and continues to make its presence increasingly known into the early twenty-first century: it is the knowledge obtained through direct *mystical experience,* a body of knowledge which has been accumulating for centuries, but only now has grown so ubiquitous that it is impossible to ignore. It is a knowledge based not on elaborate mentally produced theorems marshalled to prove the existence of God, as in the previous warfare between science and religious theology of the earlier centuries; it is based on *direct* (shall we dare say *empirical*) experience—*gnosis* repeatedly gained and described in an identical fashion by countless men and women throughout the world and among the most disparate of religious traditions. That shift in our collective understanding is happening now, one person at a time. So, wake up, my materialist friends! There is a permanent Joy within you that is awake throughout the universe and beyond. It is your true and everlasting Self. Just look with an open and surrendered heart, and you shall find it.

Notes:

1. Regarding the Big Bang and some of the modern cosmological theories, renowned mathematician and physicist, Roger Penrose, has said: "We really don't know what happened there—the big bang was a totally amazing occurrence. I don't believe any of these theories about fields we haven't found or baby universes we have no evidence for, or a larger universe in which ours is embedded. There is no objective reason to believe in any of these hypotheses. … I don't know about the cosmological constant—I don't believe in it. As for the inflationary universe theory—I am a skeptic. What these people do is come up with a theory, and when the evidence doesn't support it, they change their theory, then change it again and again." (quoted by Amir Aczel as a personal conversation with Penrose, in *God's Equation*, N.Y., Dell Publishing, 1999; pp.217-218)
2. The cyclic arising and disappearance of the universe is famously described by the mystic-author of the *Bhagavad Gita*, Chapters VIII., verses 17-20; and IX, verses 7-10. For other similar historical descriptions, see Swami Abhayananda, *Mysticism And Science*, Winchester, U.K., O Books, 2007; Chapter 8, "The Eternal Return", pp. 75-83.
3. Roger S. Jones, *Physics For The Rest of Us*, Chicago, Contemporary Books, 1992. Excerpts reprinted by permission of the author.
4. Ibid., p. 338.
5. Ibid., pp. 338-339.
6. Ibid., pp. 342-343.

2.

THE ULTIMATE THEORY OF EVERYTHING

When physicists and cosmologists talk about a 'Theory of Everything' they are referring to the potential for a theory that would provide a single unifying mathematical law governing the properties of all elementary phenomena: the various wave/particles categorized as fermions or bosons and the four known basic interactions. Such a law, if it exists, would enable these scientists to feel that they understood the means by which all the matter in the universe operates. Such a law, once formulated and proven by evidence, would be greatly celebrated among the scientific community, and would fulfill the long-sought desire on the part of physicists for a consistent theoretical framework—at least for a brief moment. For it would very quickly become apparent that there is much more to this universe than merely matter and material interactions, and that mathematical laws concerning the material universe do not answer the important questions, nor are they able to offer any lasting satisfaction in the quest for true knowledge. Such a law, if it did not take into account the Conscious eternal Source and Ruler of the universe, who constitutes the very identity of those physicists and cosmologists, would be ultimately futile and meaningless.

There can only be one ultimate theory of everything; it must be the theory that *accurately* describes the origin, evolution, sustenance, and purpose of the universe and all that's in it. And such a theory does indeed exist; it is a theory that has been both implicitly and explicitly expressed throughout the span of human history, sometimes referred to as "the perennial philosophy", but often regarded as mere myth. This ultimate theory is based entirely on direct experience, and is therefore an experientially confirmed philosophy or theory. It begins and ends with the One, known as "the Lord of the universe", "the Divine Source", "the Eternal". 'In the beginning,' this ultimate theory starts out, 'there was no universe, nor any creatures to perceive its absense; there was only the One, the "I am", who has always been. Within that One, a breath-impulse welled up, and He expelled it, projecting His own lifeforce into the simultaneously newborn spaces. And, while there were not yet any eyes to see it, it was as though a great explosion had appeared out of nowhere, from which the entire universe evolved. From Him, the universe is breathed forth; in Him it lives and evolves, and to Him it ultimately returns, in the same manner as a person's outgoing breath is indrawn once again. This world is constituted of His life's breath, and contains His life within it.

The Divine Universe

From the beginning, it is alive with Consciousness and Energy, manifesting as quanta of light and matter, and evolving into manifold forms; and this Consciousness and Energy, inherent in all matter, evolves eventually into the various sentient life-forms that populate the Earth.

All this variegated universe of form appears to exist independently as a thing in itself, with its own internal laws; but it is entirely contained in the One, consisting of His Power, and governed by His inherent and unfolding Thought. Just as men create imaginative worlds within themselves, He creates this world in time, supplying it with Consciousness and Energy out of Himself. But, just as a man dreaming is not affected by the events in his dream-world, neither is the One affected by His Mind-born creation. He remains an immaterial Presence beyond this imagined world, an eternal Consciousness in omniscient and eternal bliss. For Him, the expansion and withdrawal of this universe is but a momentary breath, though to His creatures encased in time's illusion, billions of Earth-years pass both in its expansion and in its contraction. He is beyond time and space, beyond beginnings and endings, and though He contains all things, He is uncontained, as He is the only One, besides whom there is no other.

The evolution of His cosmos brings into being sentient creatures, the most intricately evolved of these creatures being human beings. These beings inherit the eternal Consciousness of their Creator; but they also possess a false sense of individuality (called the ego), which constitutes a subtle, ideational identity (called the soul). This ego-soul comprises an ideational identity within the eternal Consciousness—which is the real underlying Identity of all human beings; and this ego-soul, in correlation with the evolving planetary patterns of this solar system, continues to evolve in intelligence and awareness through numerous lifetimes, until at last it is awakened to its true Identity. When such an ego-soul is awakened to its true Identity, it knows its true, everlasting Self as the one eternal Consciousness; and the ego-soul vanishes, as an imaginary snake disappears when it is realized to be in actuality a rope. Until such an awakening, souls continue to pass from life to life pursuing illusory selfish goals. But once having evolved, and having awakened to their true Self, such individualized souls are released from the need for further human birth, and live in the freedom and bliss of the one eternal Consciousness, serving as manifest instruments of the Divine. This is the ultimate Theory of Everything. It is discovered by each soul in its allotted time.

The empirical sciences developed by human beings serve a valuable function in that they seek to discover consistent laws governing physical phenomena, without dependence upon theoretical considerations. They seek, through pragmatic experiment and empirical sensory evidence, to derive a satisfactory understanding of universal phenomena, from the microscopic to the macroscopic, in the endeavor to formulate a consistent and accurate spectrum of human knowledge. This endeavor is both exemplary and praiseworthy; it has led to many outstanding clarifications of our understanding of the world, and has brought many improvements in the lives and circumstances of much of humanity. However, the representatives of science, by their materialistic framework and self-imposed limitation of the acceptance of empirical (physical) evidence only, have rendered science impotent to see and consider the entirety of reality, which consists of spiritual and psychological elements as well. It is as though the representatives of science have declared that *'We only deal with that part of reality that is perceivable by the senses because that is the limit of human certainty, and therefore the limit of our epistemological province; and if evidence from other experience outside that province contradicts our theories of the nature of the universe, we must simply ignore them, since such experience is not our concern'.* Thus, in their attempt to limit reality to the physical only, they have bound themselves to partial and mistaken judgments of the nature of reality. It is the task of this and future generations to correct this illogical and harmful limitation on the exploration of knowledge in all its forms, and to bring about an integral perspective that takes into account not only the physical evidence, but the psychological and spiritual evidence as well. It is only then that we will possess the capability of providing an ultimate Theory of Everything that is comprehensive, accurate, and irrefutable. Only then will the human thirst for a complete knowledge of the reality in which we live be truly satisfied.

3.

THE ORIGIN OF THE UNIVERSE

*"Every good gift and every perfect gift is from above,
and cometh down from the Father of Lights."* [1]
– from James 1:17

    The origin of the universe is a topic at the forefront of the current debate between science and gnosis; and the vision as well as the language peculiar to each of these factions makes the forging of a common understanding especially difficult. But let us try. Both the representatives of science and the representatives of gnosis agree that understanding the origin of the universe is crucial to an understanding of all that followed upon it. And I believe that the lively debate between these two perspectives is gradually centering in on a clearer presentation of the facts and a clearer understanding of the nature of the reality in which we live.

    Physicists, by extrapolating backward from the present state of the expanding universe, have theorized that what initially 'exploded' or commenced to expand in the Beginning, known as "the Big Bang", was a *singularity*. "Singularity" is a mathematical term used by physicists and cosmologists, that represents any calculation result where some property is infinite. According to calculations based on General Relativity theory, at the instant of the BB, both the density and the temperature register at infinity. So that is a *singularity*. Some physicists conjecture that one day they shall be able to eliminate these (unwanted) infinities, when a theory of quantum gravity is developed. But, for now, they are stymied. [2]

    But, from the 'spiritual' perspective, there is nothing wrong with the results of the physicist's calculations. The representatives of *gnosis* acknowledge that there was, in fact, an infinite Energy that manifested all at once along with the four dimensions of space-time. They do not dispute the notion that there was, indeed, some 15 billion earth-years ago [3], the introduction into being of an infinite burst of a spontaneously creative, consciously intelligent force that we call "Energy". But they interpret this 'singularity' as the manifestation of a Divine influx. Throughout history there have been a few persons who claim to have received, through inner vision, a spiritual revelation in which the origin of the universe was clearly shown to them. Such seers and prophets speak of the formation of the cosmos by a

Divine conscious Energy spewed forth from a transcendent Spirit (God) via His Breath, Word, or merely His Thought.

Such contemplative visions comprise some of the oldest treatments of the cosmic origin ever enunciated. We find this vision in the Vedas, in the Egyptian Pyramid texts, in the Torah, in the Bhagavad Gita, in Heraclitus, Pythagorus and Plato; and it is expressed in the declarations of mystics up to the present day. According to this 'revealed' view, there was, prior to the Beginning of Creation, no space, no time, no matter, no universe; there was only the one singular eternal Consciousness resting blissfully within Itself. And from that one Consciousness arose a creative impulse from which a breath of conscious Energy streamed forth, expanding as the universe of matter, form, and space. That (Divine) Energy had implicit in it the eternal Consciousness of its source; and it therefore had the inherent power and Intelligence to direct and self-organize itself into the form of minute particles, which then collected into more complex particles, forming rocks and stars and whole galaxies of stars; and eventually spawning the many worlds teeming with life and awareness.

It might be helpful to recall that this 'Divine Energy' is, historically, a mainstay of all religious thought. In all the world's religious theologies, the ultimate Source of all that exists is regarded as having two aspects: *the primary aspect* being the transcendent, eternal, and unchanging Consciousness (which has been called "the One", "Godhead", "Shiva", "Brahman" "Purusha", "Theos", etc.); and *the secondary aspect* is the immanent and active cosmic Energy (which has been called "*Nous*", "Creator","Shakti", "Maya", "Prakrti", "Logos", etc.). That Energy is recognized in all these traditions as an inseparable and inherent aspect of the eternal Divine. [4]

It is this Divine Consciousness/Energy which manifests as the universe of space-time and mass-energy, and which evolves as life, intelligence, and Self-awareness. *Energy* is the primordial reality. *Matter* is only the transiently stable appearance of *Energy*. Clearly, what came into being was Energy. What precipitated its expansion was Energy. Energy is the source of the universe; it constitutes the fabric and motive power of the universe; and Energy, in its various forms, is the sole existent discoverable in this universe. The only question remaining is whence comes this Energy? It is the breath of God, say those who claim to have seen into the eternal Reality.

From the perspective of contemporary physics, however, there is no actual entity such as *Energy*. It may be surprising to many to learn that the most fundamental elements of the scientific view of reality, much spoken of by physicists, both theoretical and practical, are not yet defined. Terms such as *matter* and *energy*, though utilized extensively, are beyond science's purview to define precisely. Certainly, scientists can say quite accurately what each of these abstract entities *do*, as they manifest phenomenally, but they cannot say what they *are*. But, you may protest, everyone knows what *matter* is; it's the world of substance, made up of molecules, which are made up of atoms, which are made of 'elemental' particles, which are made of energy fluctuations in the quantum vacuum. But, basically, *matter* is really the *appearance* of substance created by the activity of Energy in the environment of space-time. It is what has been called in one tradition by the name, *maya*, or illusion. But at the heart of this 'illusion' is the universal *Energy*. Energy is what burst forth as the primordial stuff, becoming the particulate *matter* of this universe. And everyone knows what Energy is – don't they?

Well, the science of thermodynamics has categorized the different 'kinds' of *energy*, such as 'potential' energy, 'kinetic' energy, 'chemical' energy, 'electrical' energy, 'nuclear' energy, and so on, according to the particular various ways energy manifests; but as for *energy* itself, there is no definition. It just *is*. We know it exists; we can describe the various ways in which it manifests, i.e., what it *does*, but we can't say what it *is*. And although we are taught in high school the rudimentary axiom that 'energy is the capacity of a physical system to perform work', which partially explains what energy does, one of the legendary physicists of our time, Richard Feynman, has acknowledged that: "It is important to realize that in physics today, we have no knowledge of what energy is." [5]

The English word, "energy", comes from the Greek, *energos*, which means "active, working", and was first coined by Thomas Young (1773-1829), a British physician, Egyptologist, and amateur physicist. And though Energy has been around forever, its first actual appearance on the scene came with a 'big bang' around fifteen billion years ago. Energy is eternal; according to the First Law of Thermodynamics, 'The Law of the Conservation of Energy: in a closed system, as this universe is, energy cannot be created or destroyed, increased or diminished. However, at its Source – that is, in the Absolute Consciousness – the primordial Energy is latent potential; it has no mass. *Mass* is defined as "the amount of inertia an object has." In the Eternal, there is no space-time, there are no objects, no gravity, and no inertia (How heavy is a rock in a dream?). *Therefore, in the transcendent Source, Energy has no mass.* Energy only becomes mass (mass-energy) upon its manifestation in space-time. *Space-time* comes into existence simultaneous with the expansion of Energy in 'the Big Bang', providing the environment for Energy to manifest as mass, hence the universe of 'matter' and 'substance'.

But, for physicists, who do not recognize *Energy* as a definable entity (aside from saying it *does* things), *matter* is yet another one of those difficult-to-define physical terms. In fact, there is no broad consensus among physicists as to the exact definition of *matter*. Usually, scientists simply avoid using the term as it is recognized to be scientifically inexact, but some use the following working definition: "Matter is any substance (??) which has mass and occupies space (although some matter has no mass, and some things may have mass without being matter)." So, while we're all comfortable that we know what is "material" and what is "immaterial", when pressed, we may have a difficult time defining what we mean in scientific terms. "Matter" may be defined only in reference to mass; and mass is measurable as energy. It is a tautological circle. Einstein has taught us that mass and energy are interconvertible; i.e., that they are simply measures of the same *thing*. Matter and energy (both of which translate as mass) clearly exist and operate, but physicists haven't a clue as to what they are. They can refer to them only in terms of what they do—in other words, by *how* they manifest in space-time.

When pressed for an explanation of the origin of the universe, some physicists state that they are able to track the manifestation of the universe back to 10^{-43} of a second after t=0 (time zero; i.e., the BB), where their calculations end in infinities. The curvature of space-time becomes infinite, curving in upon itself, and mass and energy become infinitely dense. For such physicists, this "singularity" means that, using the laws of mathematics and physics, we reach an epistemological impasse; we *cannot* know the origin, or initial state, of the universe. Otherwise, if they were to accept the singularity as a *real* objective entity, they would be faced with several unanswerable questions, like: Could an infinitesimal volume contain an infinite density? Could the entire universe fit in a pinpoint? And further, where could this pinpoint of infinite density have been located, since the universe of time and space had not yet appeared? In other words, if space was created along with time at the Big Bang, *where* was the singularity that existed before space-time came into being? Must it not be of a *transcendent* origin? Must we not concede that such an infinitely dense point outside of time and space could only be described as a latent and transcendent source of pure undifferentiated Energy? By all accounts, a *singularity*, if defined as "an infinite energy density state", is a state that is beyond the known laws of physics, since an infinite anything must, by definition, lie outside the finite universe.

But the recognition of such a 'supernatural' cause for the universal manifestation is beyond the present capability of the 'scientific' mindset; though it seems evident that the 'collective consciousness' of humanity is already openly receptive to that scenario. When physicists and cosmologists contemplate the origin of the universe, they are able to extrapolate theoretically backward to very near the beginnings of the universal expansion; but (despite the postulation of many bizarre theories) they are still at a loss to understand from whence came the initial abundance of mass-energy that precipitated and constitutes this expanding universe. Science, which utilizes the rational mind and temporal observations, is prevented from discovering the origin of the universe, since that origin is beyond both time and space and the capability of the human intellect. It is only *gnosis* which can discover the source and origin of the unverse. Unfortunately, since scientists do not believe that gnosis exists as a human possiblility, they are completely deaf to its revelations.

And so, at present, not only do scientists have no viable or even plausible theory of the origin of the universe, they remain defiantly unwilling to grant even a sidelong glance at the possibility of a supernatural cause for that origin such as has been attested to by all the seers and sages throughout history. One day, however, they will be forced by the evidence to acknowledge the primordial conscious Energy from which this universe arose, along with the supernatural intelligence and purpose inherent in that eternal Energy. And then they will come to see that, though we imagine that we are only observing this Energy manifesting all around us, that Energy has also become the ones who are watching, labeling, and interacting with itself as us. There is nothing else but that Divine Energy, and it is doing everything—including living and acting as all these self-aware human entities.

With the recognition and acknowledgement of the Divine Consciousness-Energy that originated and makes up the entire Cosmos, the undeniable evidence that this Energy consciously and independently organized itself into various forms and forces will at once become clearly and satisfactorily explained; [6] the fine-tuning of the universal conditions for life, and the actual origin (or emergence) of conscious life on earth will be satisfactorily explained; the mechanism and direction of biological evolution will be satisfactorily explained; the non-local interactions between particles of matter, and the interconnection between all conscious beings, will be satisfactorily explained; and, most importantly and significantly, the ability of human beings to know and experience, through contemplative introspection, their original Source and Identity as the one Divine Consciousness/Energy will be satisfactorily explained. With an understanding of the Divine source of Consciousness and Energy, every science (and even *gnosis*) will thus be provided with the golden key to unlock its secrets. And the answer will be forthcoming to the great question: 'If everything and everyone is a manifestation of the conscious Energy that streamed forth from God, then who am I? And who are you?'

Notes:

1. The New Testament book of *James*: 1:17.

2. There have been efforts on the part of some theoretical physicists to devise a viable naturalist scenario for the genesis of the universe that does not require a *singularity*, such as the 1973 *Nature* journal article by Edward P. Tryon entitled, "Is The Universe A Vacuum Fluctuation"; but neither he nor any others have as yet constructed a plausible theory consistent with current scientific evidence; in fact, all of the more recent physical theories designed to avoid a singularity, require the *pre-existence* of an energy-filled space (or quantum vacuum), without offering a preceding source or cause of these existants.

3. Some prefer to make this figure more exact, suggesting 13.7 billion years or 14.5 billion years ago for the Big Bang; but to avoid the frequently changing estimates, I prefer to simply say "around 15 billion years ago".

4. See my treatment of Mystical Theology in *History of Mysticism*, Atma Books, 2000.

5. Richard Feynman, "Lectures on Physics".

6. The entire universe exists as a Thought-construct in the mind of God. Therefore, all things move together of one universal accord; assent is given throughout the universe at every step of the evolutionary process: from the development and organization of atoms, to molecules, stars, planets, galaxies biological DNA, and so on, manifesting our present-day and future world. It is not necessarily a deterministic development or evolution, but nonetheless an intelligent, organic, and purposeful one.

4.

WHAT IS ENERGY?

Only one reality seems to survive ... energy, that floating, universal entity from which all emerges and into which all falls back as into an ocean; energy, the new spirit, the new god.
  – *Pére Teilhard de Chardin* [1]

Part One

When we aim to comprehend the fundamental constituency of the universe, it is useful to go back to the beginning. Physicists and cosmologists tell us that the universe began around 15 billion years ago, at the moment of "the Big Bang". And these same physicists and cosmologists surmise that what exploded in the Big Bang was a "singularity" – a superdense pocket of energy in its pre-matter state. It is often referred to as an infinitely dense "point" of zero dimensions, which, upon explosive release, became an intensely hot burst of radiant plasma rapidly expanding into and as the universe of time and space. Instantly, the *potential* energy existing in the singularity was released as *kinetic* energy in the form of an expanding fireball in whose intense heat danced the beginnings of form: quarks to construct the nucleons (protons and neutrons); and negatively charged electrons required to complete the hydrogen and helium atoms.

Later, this initially homogeneous quantum soup[2] would transform itself, from within the molten interior of stars, into heavier, more complexly organized entities, which then would be explosively dispersed throughout the universe. These elements had not existed prior to the Big Bang explosion with its consequent epiphenomena of time and space, arising synchronously with the advent of sequential events and material extension; nor had the dynamic superforce previously existed which rapidly distinguished itself into the separate and distinct forces known as the weak, strong, electromagnetic and gravitational forces. In fact, everything that now exists in and as the phenomenal 'world' was born from and is a manifestation of this initially compressed mass-energy.

This is the view of contemporary physicists and cosmologists, who regard the primal Energy as the single source, substratum and lifeforce of the universe, pervading even the vacuum of deep space. However, it sometimes happens that the language that we use to talk about things lags behind our intuitional understanding of those things. And I believe such is the case regarding the application of nineteenth century language to our twenty-first century understanding of energy. When you examine the textbooks of the scientists in search of their insights regarding this fundamental ocean of unfathomable power which we call *energy*, you discover that their definition of the word has barely changed since it was originally stated: "*Energy* is the capacity of a body or system to do work, or to release heat or radiation." This is but a slightly modified version of the definition produced in the nineteenth century, when scientists were first quantifying and measuring the energy and work beginning to be produced by the industrial revolution. *Energy is* the capacity for work, to be sure; but that, it seems to me, is the least of energy's characteristics. When we say, "Prior to the Big Bang, from which space, time, and matter originated, there was only *energy*", we are using the word, *energy*, to represent a distinct entity in itself, a seething cauldron of some actual thing, some inconceivably immense and indefatiguably creative power. Certainly, *energy* exists as a capacity, a potential; but it also represents in its larger meaning the core fountainhead from which flows the substance and essence of the entire phenomenal world of forms.

Energy is not a blind and lifeless force; Energy is replete with divine Intelligence. It has created this entire universe by itself with no diagrams or plans, but assuredly goes about its task of building this universe in accord with its own divinely inspired design. First, it parcels itself out into tiny discreet entities or *quanta*; then these automatically attract the appropriate elements required to make more complete constituents of yet larger and more complex structures called *atoms*. These, in turn, of their own proclivity, combine to form larger clusters called *molecules*. By its own internal blueprint, Energy constructs the immense world of myriad objects and organisms, eventually culminating in mankind, imbuing him with life and strength and wisdom. What an incredibly amazing thing is this Energy! And yet we belittlingly refer to it as "the capacity for work"!

Nor does one find, in any of the reference books on physics or cosmology, the slightest reference to this larger definition of energy, revealing it as the cosmic sea of activity out of which all phenomenal appearance is made. It was Einstein who familiarized us with the understanding that *mass* and *energy* are interconvertible, as shown in his formula, $E=mc^2$; and it is now commonplace to refer to an object such as a wave/particle in terms of its mass/energy. We have learned, in fact, that mass (and, by extension, matter) and energy are not only interconvertible; they are the same *thing*. But with the present (archaic) definition of *energy*, Einstein's formula merely informs us that mass times the speed of light squared is equal to a certain amount of *work capacity*. In the real world, however, physicists as well as metaphysicists often refer to "Energy" in its larger meaning, as the source and substance, creator and architect, of all that exists in this universe. So it is clear that there is a common usage of this word, *Energy*, that is not being clearly defined or acknowledged.

Is it possible that we need a new definition of *energy* to match the actual way that we use the word? Here's one that might fit the bill (notice that this definition requires that the first letter of the word be capitalized):

"Energy is the elemental creative force, responsible for the manifestation and proliferation of universal phenomena, including matter, motion, force, heat and radiation." [3]

And, while we're at it, let's redefine *matter* as well: *"Matter is Energy manifested as distinctly discreet phenomena subject to the attraction or repulsion of the elemental forces."*

Now that we have established the meaning of the word *Energy*, we can talk meaningfully about the conversion of Energy to matter, and matter to Energy; and we can more comfortably talk about Energy as the fundamental constituent of universal manifestation. We can even investigate the invisible Source of Energy with a greater possibility of reaching agreed upon conclusions about its transcendent origin.

Part Two

Let us go back now, for a moment, to what the physicists call the "singularity" from which all this Energy/matter erupted. It was first surmised that such a thing as a singularity existed when it was discovered by astronomer Edwin Hubble that the universe was expanding. It must, reason suggests, have been expanding from a previously more contracted and dense state. Following this line of reasoning backward, there had to have been an initial state of infinitely contracted, infinitely dense matter; and since matter at such a contracted and dense state would have beeen super-hot, it would have been converted into pure Energy; and at some point this Energy exploded as what we now call "the Big Bang". This theory received convincing confirmation when the lingering background microwave energy (electromagnetic waves) from that initial explosion was detected by radio astronomers, Penzias and Wilson. Okay, now we had an infinitely dense and dimensionless point, calculated to have existed around 15 billion earth-years ago, from which this currently vast universe burst forth. But we can't leave it simply at that; we must question "Where did this singularity come from?"

I would like to put forward the notion that there really was no *singularity* at all, unless we mean by the word, "singularity", the One without a second, the Divine Ground, the Father of all. For it was He, that imperceivable Energy-producing Spirit, eternal and transcendent, who, by His own power of radiant "emanation", breathed into manifestation the explosion of Energy that became the time, space, and material substance of this phenomenal universe. This is not a theoretical concept produced by the faculty of reason; it is what I have distinctly seen in the depths of contemplative vision. I believe that it is in keeping with the contemplative vision of countless other seers who preceded me. In any case, for me it is a certainty based on direct revelation, a revelation and a certainty which is indelibly imprinted in my mind. The Source of this breathing forth I cannot describe; I can only say that It is a solitary eternal Consciousness that is both distinct from and integral to the universe it projects within Itself. The closest I am able to come to an analogy is that of the projection of an elaborate dream in the mind of the dreamer of that dream. Just as a dreamer's mind exists apart from the dream and yet provides the consciousness and animating power of the dream, so does that one Spirit/Consciousness exist in His own manner, distinct from this elaborate dream-universe, and is yet at the same time the very substance of this universe and all that is in it.

For Him, the explosive appearance of universal manifestation that we call "the Big Bang" is merely the instantaneous initiation of an Energy-laden breath; and the collapse and return of all the matter/Energy to its initial potential state is but the alternate cycle of that breath. From His perspective, the expansion of the universe and its subsequent contraction is a momentary event, a mere breath—followed by another breath, and another, without end. In each of these breaths, the many worlds are manifested in which inumerable souls pass through the evolutionary course of their adventure in these dream-worlds toward a clarified awareness of the one transcendent Consciousness from which they are born and in which they live and move and have their being. In the next breath following, the many souls breathed forth by the One Soul will pass through that long course once again.

It is important to note that each soul, while it has a semblance of individuality, is in essence the one Consciousness, each one being identical to its Source, just as each piece of earthenware, despite its individual shape and form, is identical to the clay from which it is made; or, more analogously, as each dream-character in a dream may trace its consciousness to the one consciousness of the dreamer. Notice that, in the mind of a dreamer, we may distinguish between the background consciousness and the form-producing imagination. The dreamer's consciousness provides the identity, the I-awareness within the dream; while the image-making faculty, the projecting power of the mind, provides the animation, the pictures. And while the two—image-making faculty and background consciousness—retain a semblance of distinction, they exist together in the one mind. This is a clue, hidden in our own existence, of the manner of God's projection of this universe. For, just as a dream-character's form is made from the same substance as its awareness, so also we, in this dream-like reality of time and space are not divided in regard to our body and soul, or mind. There is an appearance of separation, to be sure; but the fact is they are made of one and the same Consciousness, and live within that one Consciousness as an integral whole.

There is no need, therefore, to join with God; we are already one with Him. What must be attained is the *knowledge*, the lasting awareness, of our Divine identity. The destination of each soul is the *recognition* that it is in truth the transcendent Spirit from which the entire universe emanates. It sees within its own soul-vision the emanation of the many worlds from its own Self, and realizes that it is, Itself, the one universe-breathing Spirit, eternal and unchanging, alone and untroubled, blissfully including all within Itself. And that soul then knows it has reached the summit of its searching, and the fulfillment of its impelling desire. All is clear; there is only Itself, amusing Itself in the splaying forth of multiplicity in universe after universe. *Nothing to lament, nothing to pride oneself on. All is accomplished in an instant. ...I have but breathed, and everything is rearranged and set in order once again. A million worlds begin and end in every breath, and in this breathing, all things are sustained.*

Notes:

1. Teilhard de Chardin, *The Phenomenon of Man*, trans. By Bernard Wall, N.Y., Harper, 1959; p. 258.

2. George Gamow, one of the early 'Big Bang' theorists, referred to this primordial soup of Energy of which the universe is composed as *ylem* (pronounced "í-lem"), an archaic word which Webster's Dictionary defines as: "the primordial substance from which all the elements have been derived."

3. Nearly every Physics textbook I looked at merely stated the old chestnut: "Energy is the capacity for work." Even Alan Guth of Inflationary fame repeated it, saying: "Roughly speaking, energy is the capacity of any system to do work." (Alan Guth, *The Inflationary Universe*, N.Y., Addison-Wesley Pubs., 1997; p. 3).

Some mildly objected to this definition, modifying it somewhat, such as this: "Energy [is] the measure of the ability of an object (whether a photon of light, a tennis ball or an entire galaxy) to affect another object. It is often defined as the object's capacity for work, but this definition fails to take account of entropy (heat energy unavailable to perform work)." (Christopher Joseph (ed.) *A Measure of Everything*, Buffalo, N.Y., Firefly Books, 2005; p. 138). I have found only one theoretical physicist who acknowledges the broader definition of "Energy". Here, from his Internet Encyclopedia of Science, intuitive and innovative physicist and astronomer, David Darling, Ph.D., gives both the traditional and modern definitions of energy: "1. A measure of the ability to do work – for example, to lift a body against gravity or drag it against friction> or to accelerate an object. 2. An intrinsic property of everything in the universe, including radiation, matter, and, strangely enough, even empty space." (www.daviddarling.info), For an interesting and satisfying reconciliation of these two definitions of energy, see Appendix 1: "Reflections on The Two Definitions of Energy" at the end of the book.

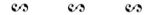

5.

WHERE CONSCIOUSNESS COMES FROM

> *"Consciousness, in order to be integrated into a world system, necessitates consideration of the existence of a new aspect or dimension in the stuff of the universe."*
> *— Pierre Teilhard de Chardin* [1]

No idealogical conflict better represents the idealist-materialist divide than that between the materialist Neurobiologists who claim that human consciousness is a product of neural activity in the brain, and those idealists who assert the primacy of Mind and Consciousness as the source and substance of the universal creative force of which matter (including brains) is constituted. The idealistic position goes back thousands of years, and is reflected in the various religious views of the origin of the cosmos, and in the Platonist tradition as well. That position was reiterated in the philosophical view of René Descartes (1596-1650), who asserted that mind (spirit) and matter were two separate *kinds* of existents comprising man— both emanating from God, but with differing characteristics. This was the basis of the well known philosophy of Cartesian dualism, which holds that these two categories are inviolably separate and distinct entities: one, the Divine uncreated part of man (the spirit); the other, the created form-manifesting part (the body). Though this philosophy offered no essential modification to earlier Platonist thought, it was the product of a rational introspection that proved appealing and persuasive to many of its time. The scientific materialism of the nineteenth century found no place, however, for the soul, and presumed to repair this conceptual mind-body split with the belief (still current) that all that is is solely material, including mind; and that such a thing as 'spirit' or 'soul' does not exist.

For contemporary materialist science, there is no God, no soul, and mind is merely a manifestation of the activities of neurons and synapses in the brain. In describing the origin of the cosmos, today's materialistic scientists start with the assumption of the existence of a 'singularity', wherein an infinitely dense mass of plasmic energy is crammed into an infinitesimally minute speck of potentiality. Then, due to some random fluctuations, it bursts its bounds, exploding outwardly (away from its center) to become the expanding universe of space, time, matter and invisible forces. This is the theoretical picture that the currently accumulated scientific evidence paints. Scientists do not even question what produced this singularity; i.e., why there is something rather than nothing, and how it happened to be. Furthermore, these materialistically inclined scientists are placed by this theory in the uncomfortable position of being required to explain how human consciousness emerged or evolved from the cooled remains of this boiling soup of primal plasmic energy.

Currently, in the early part of this twenty-first century, scientists—Physicists, Cosmologists, and Neurophysicists—are busily pursuing the theory in which consciousness, that most evident of existents, somehow arose some 300,000 years ago as an 'epiphenomenon' of the self-organizing activity of brain cells and neurons; i.e., just popped out of biological tissue by some as yet unknown process of spontaneous production, and is basically a phenomenon arising from the activity of biological matter. Here is a statement of that theory by a contemporary professor of philosophy: he states that consciousness

> is a biological feature of the human and certain animal brains. It is caused by neurological processes and is as much a part of the natural biological order as any other biological feature. [2]

Others, more cautious, say merely that

> Consciousness indubitably exists, and it is connected to the brain in some intelligible way, but the nature of this connection necessarily eludes us. [3]

Another says:

> I doubt we will ever be able to show that *consciousness is a logically necessary accompaniment to any material process*, however complex. The most that we can ever hope to show is that, empirically, processes of a certain kind and complexity appear to have it. [4]

Over the years leading up to the present (2008 C.E.), little progress has been made in the attempt to formulate a satisfactory theory of the material origin of consciousness. In the beginning of a recent book of memoirs (2006) by Nobel prize-winning Neurobiologist, Erich Kandel, a hopeful and promising picture of future progress is offered:

> The new biology of mind ...posits that consciousness is a biological process that will eventually be explained in terms of molecular

> signaling pathways used by interacting populations of nerve cells. ... The new science of mind attempts to penetrate the mystery of consciousness, including the ultimate mystery: how each person's brain creates the consciousness of a unique self and the sense of free will. [5]

But then, in the latter part of the book, he admits that

> Understanding Consciousness is by far the most challenging task confronting science. ...Some scientists and philosophers of mind continue to find consciousness so inscrutable that they fear it can never be explained in physical terms. [6]
> What we do not understand is the hard problem of consciousness—the mystery of how neural activity gives rise to subjective experience. [7] ...Biological science can readily explain how the properties of a particular type of matter arise from the objective properties of the molecules of which it is made. What science lacks are rules for explaining how *subjective* properties (consciousness) arise from the properties of objects (interconnected nerve cells). [8]

As I have stated repeatedly in the past, this search is a misguided one, and can only lead to a dead end; for consciousness does not arise from neural activity in the brain; it is a primary property of the Divine Mind, and is implicit in the universal manifestation, and therefore in all matter. It is what matter is made of. However offensive such a statement may be to others, I must declare that this truth has been revealed to me; I've *seen* it. I cannot, therefore, present this truth as a rationally fashioned postulate; it is a directly perceived *fact* that has been clearly revealed in the unitive vision not only to myself but to many others who have been graced with that vision. To understand the truth of the origin and constituency of this universe requires an uncommon perception which comprehends all forms in the universe as manifestations of the one transcendent Consciousness. The eventual acknowledgement of this truth will require a radical transformation in the thinking of all men and women of science which, though it may take centuries in which to unfold, will usher in a truly golden era of Enlightenment.

Today, we look back on the contemporaries of Copernicus with the advantage of hindsight, and wonder how the intelligentsia of that time could possibly have failed to perceive that the earth travels about the sun, and not vice versa. Once the truth is known, the errors of the past seem so obviously unsupportable. Once the light shines, the preceding darkness is clearly recognized. One day, when it is readily recognized and acknowledged that the world of space, time, matter and energy arise from the Divine Consciousness, men will wonder how it could possibly be that once seemingly intelligent people thought that consciousness was an epiphenomenal product of biological matter.

Consciousness is, in fact, beyond time and space, and all manifestation; It is the eternal Identity of all that exists. It transcends the universe, while constituting its essence—as a dreaming mind transcends its dream-images, while constituting their essence. Consciousness is not the property of matter, or of any individual being. It is not produced by any material process; but rather is the producer of all matter and all processes. It is the fundamental nature of Being, the foundation of the phenomenal universe, the Light of the Projector which flashes its images in the space-time dimension which we know as 'the world'. The projected human images on this screen can only know that eternal Consciousness by following their own consciousness back to its Source, where they will discover their own Divine Self.

There they will discover that their own consciousness is the one eternal Consciousness that is the sole Being. The phenomenal universe, said by scientists to be made of many different particles of energy to which they have given many names, is in reality made of the Divine Consciousness. We may find a clue to understanding this by pondering the nature of our own minds, since, as has often been said, we are images of God. Consider the nature of our dreams: the consciousness of the dream-character is really the consciousness of the dreamer, is it not? And what of the body of the dream-character? Is it not a projected image produced by the dreamer's mind, and consisting also of consciousness? By analyzing this clear analogy, one may begin to have a notion of the irrelevance of Descartes' conception and also of that of the materialist scientists. But, of course, one must see it for oneself. One's mind must be illumined by the eternal Light itself, and drawn into Its hidden depths. To obtain that grace, all men focus their minds on Him through prayer and contemplative longing, and He shines His Light on whom He will.

Notes:

1. Pierre Teilhard de Chardin, The Phenomenon of Man, trans. by Bernard Wall, N.Y., Harper, 1959; p. 55.

2. John Searle, professor of philosophy at U.C. Berkeley, quoted by Richard Restak, *Mysteries of the Mind*, Washington D.C., National Geographic, 2000; pp. 71-72.

3. Colin McGinn, *The Mysterious Flame,* quoted in R. Restak; *Ibid.;* p. 85.

4. Jeffrey Satinover, *The Quantum Brain,* N.Y., John Wiley & Sons, 2001; p. 220.

5. Eric R. Kandel, *In Search of Memory: The Emergence of a New Science of Mind,* N.Y., W.W. Norton & Co., 2006; pp. 9-11.

6. Kandel, *Ibid.;* p. 377.

7. Kandel, *Ibid.;* p. 382.

8. Kandel, *Ibid.; p. 381.*

6.

TIME, ETERNITY, AND THE FUTURE TASK OF SCIENCE

Part One:

Time And Eternity

Newton believed in an absolute time; one which is always the same for everyone in every situation. Einstein demolished that view by showing that the measure of the passage of time is relative to motion—differing by the variation in motion between two perceivers. Cosmologist Stephen Hawking further clarified time's non-absolute status by noting that "time is just a coordinate that labels events in the universe; it does not have any meaning outside the spacetime manifold."[1] Indeed, space and time (space-time) only come into existence along with the birth of the universe. Cosmologists assert that, prior to the genesis of time and space, there was only a "singularity", a mathematical point of concentrated energy possessing zero volume and infinite mass from which the universe explosively expanded. In that instant when that energy let loose as the "Big Bang" and began to expand as the plasma that would become particulate matter, space and time also came into existence. Before that, space-time did not exist. To the question, "What was when space-time was not?", the answer is, "Eternity".

Now, from a purely theoretical point of view, Eternity can be a very daunting concept, one which cosmologists as a rule refrain from considering. But for those of us who have been privileged to *experience* Eternity directly, it is neither a theory nor a concept. We know, with absolute certainty, that Eternity is the underlying foundation, support, and projecting power upon which this universe of time and space exists. We know that time exists only in the universal manifestation, with a recurrent beginning and end, and that in Eternity there is no such thing as time – no past, no present, no future, no projected universe at all. For Eternity is just another name for the absolute Mind that is the ground and support of the universal projection; and it is the source of the consciousness which sentient beings experience within themselves. Eternity is the upper reach of Existence, to which the mind may be drawn, if God so wills; and there it is seen that time has no absolute existence, but exists only as one of the dimensions, along with space, of the Divinely projected imagination in the Mind of God called 'the universe'.

The Divine Universe

This universal manifestation is superimposed upon Eternity, as a dream is superimposed upon the consciousness of a dreamer. One could say that the temporal universe and Eternity exist in separate levels, or gradations of Consciousness – as the dreamer and his dream-world exist in separate levels, or gradations of consciousness. Eternity is at the highest level of consciousness. It is experienced by a mind that is intensely and utterly focused and intent upon the Divine. It completely supplants one's limited individuality, raising one's awareness to Its own place, and revealing one's ultimate identity with Itself. This experience of Eternity is quite pleasant. It is impartible, single; perfect aloneness, blissfully content. It sends forth a new universe in every breath, while in the same alternating breath annihilating the old. It is so simple and unencumbered that it cannot be conveyed in speech. It is the ancient, unnamed God. It occupies its own place, its own dimension, quite sovereign and alone. The temporal array spewed out in each breath offers no distraction or interruption to the sweetness of Its homogeneous peace. It is its own perpetual delight and satisfaction. The cosmos, quite a different thing, originates from Him, and dissolves in Him; and time derives from Him, though He is utterly beyond time's reach. It is as a dream, emanating from the mind of a dreamer, exists in its own place, depicting a drama, originating, then reaching a culmination, but in no way affecting the dreamer; even though each of the dream characters is, in reality, the dreamer, who, once awakened, returns to the awareness of its true source and Self.

The same scenario takes place in this projected 'real' universe of time and extension. We who live within it are all none other than the one Eternal Mind, and on awaking shall once again know our blissfully eternal Self. And even now, in this temporal moment, in this spatial unfoldment of the cosmic dream superimposed upon the eternal Consciousness, we are in truth that one eternal Self, blissfully content, fully awake, in our solitary timeless, spaceless place on high. And while this imaged time, begun in that first instant of cosmic appearance along with space, marches on, we momentary creatures move to its rhythms without knowing why or whence, yet happily knowing, by the creator's grace, our everlasting Self beyond time, and singing praise and glory to His name.

Notes:
1. Stephen Hawking & W. Isreal (eds.), *300 Years of Gravitation*, Cambridge University Press, 1989; p. 651; quoted in P. Coveney and R. Highfield, *The Arrow of Time*, N.Y., Ballantine Books, 1991; p. 99.

Part Two:
The Future Task of Science

While the understanding described above is a perfectly clear and satisfying worldview for the mystic, for the scientist it brings up at least one important question; it is the paramount question: '*Where is the empirical proof that the eternal Consciousness transforms into the Energy that was made manifest in the Big Bang?* It is a question for which the mystic has no satisfactory answer. Scientists already know that Energy and mass (matter) are interchangeable, and so the understanding of the transformation of the intense concentration of Energy into the wave/particles of matter that makes up our universe has already been scientifically established. But, it is the as yet undiscovered evidence for the *initial* transformation from the eternal and noumenal Consciousness to that phenomenal space-time Energy that is the sticking point, the stumbling block, to a complete and comprehensive understanding of the means by which the Divine Ground projects the appearance of universal manifestation. The means for the transition from Eternity to time, from Consciousness to Energy, from the Transcendent to the immanent, cannot presently be demonstrated or even rationally explained.

To illustrate this difficulty, let us go back to that initial appearance of Energy, the so-called "singularity" and its bursting forth as the "Big Bang". We know what happened when that singularity (a volumeless point of infinite mass) released its potency into the actuality of space-time; 'But what about prior to that?' we want to ask; but alas, there was no "prior", since time had not yet been born. 'So, where did it come from?' we start to ask; but that too is unaskable, since there was no "where" prior to the birth of space. The question then becomes 'How did the eternal Consciousness manifest as the singularity?' The answer that might be given is that there was no singularity other than the eternal One. It was He who manifested the abundance of Energy which burst forth as the universe of form. This universe was manifested in the mind of God, just as a thought-form is manifested in the mind of man. This is the answer that comes from the vision of the seers, the mystics; but in order to be accepted by all, it must somehow be confirmed and fleshed out in a consistent, empirically verifiable, 'scientific' theory.

From my own and others' mystical vision into the Eternal, it may be confidently declared that it is the one eternal Consciousness who projects, emanates, or otherwise sends forth out of Itself this universe of appearance; and we affirm that that inconceivably immense quantity of Energy from which the universe is born comes from the 'dimension' of Eternity. It simply bursts into being as the phenomenal universe from that eternal Source outside the spacetime manifold, as the appearance of an immense landscape might burst into being in a dream originating from the creative unconscious of a human being. How does a dream-figure take form in the mind of a human being? Perhaps if we could understand *that* process, we might gain some insight into the way the universal Consciousness takes form as the energetic wave/particles of the phenomenal universe. We wish to create a beautiful young woman in our imagination, and immediately there she is! Her appearance involves nothing more than a slight shift of concentrative focus, and *voila*! That creation is instantaneously produced in the mind's eye by an effortless act of will. It is suggested that God's will operates in a comparable manner to produce a universe that bursts into being in a spectacular fashion.

If science is to put its attention to bear in a serious investigation of these tenets, it must find a way to uncover the presence of Consciousness within the structure of matter. It seems to me that the great mystery to be solved is *the question of how God's Thought-Energy becomes the wave/particles of which atoms, molecules, etc. are formed!* By what inscrutable alchemy does insubstantial (Divine) Thought-Energy transform itself into discreet wave/particles which in turn become substantial forms? Therein lies the mystery of the universe, the riddle to be solved. *Here, then, is the future task of science:* while you scientists have already shown that the fluctuations in the vacuum of space-time show up in the form of wave/particles electrons, photons, and quarks; now you must prove that the underlying nature of Energy and space-time is Consciousness. Abandoning the search for a material basis for consciousness, you must search for the Consciousness that is the basis for matter. You must discover that Consciousness within mass/energy; and prove the theorem that states: Consciousness=mass-energy ($C=me$).

When this is accomplished, then we can be rid at last of the distasteful and unsatisfactory *materialism* that has beleaguered us for so long, and which has been such a crippling shackle on our understanding. *Then*, and only then, will science have fulfilled its centuries-long God-given task of producing a truly Grand Unified Theory of Everything. Then, the artificial Cartesian separation of mind and matter will be destroyed forever; and science and spirituality will be united in a common nest of Truth, sharing a common view of all aspects of the Divine reality in which we live. Then will time truly bear its latent fruit, when all people join in knowledge and in love, conscious of their common birth from the heart of their eternal Father, and united in their aspiration to fulfill their Divine birthright as the Sons and Daughters of the bountiful and eternally gracious God. Scientists of the world, we're counting on you!

☙ ☙ ☙

7.

IN THE FINAL ANALYSIS

Part One

We can imagine some ancient archetype of man looking out at the world around him, and wondering to himself, "What is this world I find myself in? What is it made of?" The next question would no doubt have been: "Who made it?" Over the ages, as men have continued to ask these questions, different answers have emerged, but with little consensus of opinion. Let's review and see if we can sum up where we are today in the quest for an understanding of the world of appearance that we call "reality".

First of all, we have to acknowledge that there has been, since the beginning, two basically different approaches to these questions: the one approach is intuitional; the other is experimental. The intuitional approach seemed at first to offer common sense answers to these questions, but these were later shown by the experimentalists to be either false or untestable. And the experimentalists, often taking only the outer appearance of phenomena for the entirety of reality came frequently to conclusions which failed to account for the invisible subtleties of that reality, such as consciousness, dreams, or conceptions. But, as time went on, with both of these factions *working together*, they eventually formulated a working hypothesis that seemed to explain everything that makes up our "reality".

Beginning around the sixteenth century of our Current Era, a few men began to hone in on the basic laws and constituents of the reality in which they lived, both on the intuitive and on the experimental side. And now, after many revisions, mankind as a whole, through both the intuitive method and the experimental method, has finally reached a consensus in the early 21st century regarding the answer to the questions of what this world is made of and who made it. Here is a brief synopsis of that accumulated consensus:

The experimentalists discovered that all phenomena perceived by the senses could be reduced to the category of "mass-energy"; and this mass-energy appeared to arise spontaneously from the single infinite unified field called "space-time", and could be seen as either particulate or wave-like, depending on how it was measured by human consciousness. Therefore, the experimentalists, through their investigations into the nature of material reality, have determined that the world (universe) is made of the "consciousness of space-time's universal mass-energy" (or COSTUME).

The intuitionalists have, in turn, determined that the COSTUME may be seen to possess four different layers, each one subtler than the previous one. These are, in descending order: the gross, astral, causal, and eternal (or GRACE). These levels are like the layers of an onion, one more interior than the other; but they exist simultaneously. The Eternal level, experienced in mystical vision as the transcendent Consciousness, is the *Ultimate Source* (or US); the Causal level is the initial projection of the creative Power of the One, also perceivable from the contemplative state; the astral level is synonymous with the soul level, where intricate patterns of individual karma may be seen to reside; and this astral level interacts with and gives expression to the gross level that is apparent to the senses. These various "orders" of reality reside within each other, each with its own reality, but each interacting with and feeding into the other. Thus, the eternal Consciousness that is the primary principle and Ultimate Source (US) continues to inhere implicitly as consciousness in each subsequent level or order of the COSTUME (For more about this, see my book, *Mysticism and Science*).

The intuitionalists hold that GRACE also holds the answer to the question of "Who made this world", as all is clearly reduced to the Eternal (US), as the beginning and causal impetus of the world of appearance we call 'reality'. The Eternal is to be considered the repository of all causation: the material cause (i.e., what it's made of), the efficient cause (i.e., the immediate effective power), the formal cause (i.e., the form or design followed), and the final cause (i.e., the purpose or motive guiding the production) of all that is. Thus, in the final analysis, it is determined that the intelligible universe is the COSTUME; the COSTUME is made of GRACE; and GRACE is produced by the ETERNAL, or US. And so it is clearly apparent that, now in our present space-time, all the questions of our archetypal man have been answered satisfactorily, and, thus enlightened, all beings may now live happily ever after.

Part Two

"Wait a minute", you're saying to yourself; "even if that was an accurate account of reality, and I accepted it fully, that wouldn't necessarily make me enlightened and happy ever after!" And you're right. Even if we could conceptualize every facet of reality, and comprehend everything about everything, we would still not necessarily be enlightened, nor would we obtain lasting happiness. Neither scientific nor philosophical knowledge is able to supply us with the perfect satisfaction of enlightenment. For that, we need to wake up.

The more we know of this dream-like world in which we live, the more clearly we recognize its inadequacy to bring us the satisfaction we crave. We may indeed possess a full appreciation of the beauty, intricacy and incredible diversity of God's amazing world, and yet, again and again, we are brought back to the recognition that it is all mere ephemera, mere images; it is the incredible play of intangible yet energetic "forces" all together weaving this magic show for our delight and wonderment. We realize that, though we participate in the drama, we are not imprisoned in it. We are something different; we are the Mind, unmoved and unchanging, from whom this drama emanates and unfolds. We are the universal Delight that imagines this world of light and sorrow, hope and decay, and, in order to know that Self, we simply need to wake up.

Of course, it is not *really* a dream we find ourselves in! Yet what better analogy can there be to this world than that of a dream? Having once awakened, I know my eternal Self to be the Dreamer: the producer, witness and enjoyer of the dream. And while there is certainly much to understand about "the dream", its mechanisms and propensities, the truth is it is all mere image. The Real is the eternal 'I' that periodically projects this fantasy on its own screen. And that is what we need to concern ourselves with; what we must revere; and what, with all our heart and soul, we must strive to realize.

The perfect satisfaction you crave can only come by 'waking up' from the false illusory "I" to know your true eternal Self. By "knowing your eternal Self" I mean the ascension of your limited consciousness to its original and unlimited source. I mean the clear awareness of God as the true 'I' behind all "I"s. I mean complete absorption in Him. Of course, there is no "Him"; that's just the way we speak while we're in the 'dream'. Both the dream "I" and the dream "Him" disappear when you awake to your true Self. Then, there is only the One.

Unfortunately, however, you cannot awaken yourself just because you've been told that you're dreaming and need to wake up. "He" must wake you. Still, when you feel a strong desire to know your true Self, it is a good

The Divine Universe

indication that you're being pulled from within toward that awakening, and that you should set the stage for inward reflection. The awakening will occur in its own time. Having once experienced it, I know of its possibility; and having once experienced it, there is nothing else I am capable of desiring. Those who take satisfaction in mere philosophy, in mere words, cannot understand the heartache of those who seek His touch, who yearn to be enfolded in His embrace. This heartache is a holy sickness that He imparts to those He wishes to awaken. Retain that sickness, and nourish it; it will draw you home. There only will you find enlightenment and happiness ever after. But remember, *you* cannot accomplish it! As a great awakened sage once said, "For man it is impossible; but, for God, all things are possible."

II.

SPIRITUAL VISION

When we study the many speculative philosophies and religious creeds which men have espoused, we must wonder at the amazing diversity of opinions expressed regarding the nature of reality; but when we examine the testimonies of the mystics of past and present, we are struck by the unanimity of agreement between them all. Their methods may vary, but their ultimate realizations are identical in content. They tell us of a supramental experience, obtained through contemplation, which directly reveals the Truth, the ultimate, the final, Truth of all existence. It is this experience, which is the hallmark of the mystic; it goes by different names, but the experience is the same for all.

By many of the Christian tradition, this experience is referred to as "the vision of God"; yet it must be stated that such a vision is not really a "vision" at all in the sense in which we use the word to mean the perception of some 'thing' extraneous to ourselves. Nothing 'other' is at all perceived in "the vision of God"; rather, it is a sudden expansion, or delimitation, of one's own awareness which experiences itself as the ultimate Ground, the primal Source and Godhead of all being. In that "vision," all existence is experienced as Identity.

We first hear of this extraordinary revelation from the authors of the Upanishads, who lived over three thousand years ago: "I have known that spirit," said Svetasvatara, "who is infinite and in all, who is ever-one, beyond time." "He can be seen indivisible in the silence of contemplation," said the author of the Mundaka Upanishad. "There a man possesses everything; for he is one with the ONE." About five hundred years later, another, a young prince named Siddhartha, who was to become known as the Buddha, the enlightened one, sat communing inwardly in the forest, when suddenly, as though a veil had been lifted, his mind became infinite and all-encompassing: "I have seen the Truth!" he exclaimed; "I am the Father of the world, sprung from myself!" And again, after the passage of another five hundred years, another young man, a Jew, named Jesus, of Nazareth, sat in a solitary place among the desert cliffs of Galilee, communing inwardly, when suddenly he realized that the Father in heaven to whom he had been praying was his very own Self; that he was, himself, the sole Spirit pervading the universe; "I and the Father are one!" he said.

Throughout history, this extraordinary experience of unity has repeatedly occurred; in India, in Rome, in Persia, in Amsterdam, in China, devout young men and women, reflecting on the truth of their own existence, experienced this amazing transcendence of the mind, and announced to everyone who would listen that they had realized the truth of man and the universe, that they had known their own Self, and known it to be the All, the Eternal. And throughout succeeding ages, these announcements were echoed by others who had experienced the same realization: "I am the Truth!" exclaimed the Muslim, al-Hallaj; "My Me is God, nor do I recognize any other Me except my God Himself," said the Christian saint, Catherine of Genoa. And Rumi, Jnaneshvar, Milarepa, Kabir and Basho from the East, and Eckhart, Boehme and Emerson from the West, said the same.

These assertions by the great mystics of the world were not made as mere philosophical speculations; they were based on experience–an experience so convincing, so real, that all those to whom it has occurred testify unanimously that it is the unmistakable realization of the ultimate Truth of existence. In this experience, called samadhi by the Hindus, nirvana by the Buddhists, fana by the Muslims, and "the mystic union" by Christians, the consciousness of the individual suddenly becomes the consciousness of the entire vast universe. All previous sense of duality is swallowed up in an awareness of indivisible unity. The man who previously regarded himself as an individualized soul, encumbered with sins and inhabiting a body, now realizes that he is, truly, the one Consciousness; that it is he, himself, who is manifesting as all souls and all bodies, while yet remaining completely unaffected by the unfolding drama of the multiform universe.

Even if, before, as a soul, he sought union with his God, now, there is no longer a soul/God relationship. He, himself, he now realizes, is the one Existence in whom there is neither a soul nor a God, but only the one Self, within whom this "imaginary" relationship of soul and God manifested. For him, there is no more relationship, but only the eternal and all-inclusive I AM. Not surprisingly, this illuminating knowledge of an underlying 'I' that is the Soul of the entire universe has a profoundly transformative effect upon the mind of those who have experienced it. The sense of being bound and limited to an individual body and mind, set in time and rimmed by birth and death, is entirely displaced by the keenly experienced awareness of unlimited Being; of an infinitely larger, unqualified Self beyond birth and death. It is an experience which uniquely and utterly transforms one's sense of identity, and initiates a permanently acquired freedom from all doubt, from all fear, from all insecurity forevermore. If we can believe these men, it is this experience of unity, which is the ultimate goal of all knowledge, of all worldly endeavor; the summit of human attainment, which all men, knowingly or unknowingly, pursue.

While it is held by many deep thinkers and philosophers that the existence of God is not subject to empirical proof, in fact, it is. The empirical "proof" of God is the direct experiential vision of God, just as the empirical proof of the Sun is the direct experiential vision of the Sun. In the West, that is, in the Platonic and Judeao-Christian traditions, this unitive vision is referred to as "mystical experience". And yet, to many, "mysticism" implies mistiness, vagueness, and all that is mysterious. Among materialistic scientists, it is a term of derision; for them, it implies mystification, fuzzy-mindedness and delusion. Nonetheless, for centuries, "mysticism" has remained the only label applicable in the West to the claims of unitive vision; but since it has so consistently aroused the distrust of the more common variety of men — materialists and 'pragmatic realists' all — perhaps the term, "mysticism", should be replaced with the term, "empirical religion"; for the mystics of the world are truly the empiricists of religion.

Whatever the term we may ultimately use, I am focused, in this second grouping of Essays, on the visionary revelation commonly known as "mystical experience", or "enlightenment". While it cannot be produced volitionally, it is a gift of God that visits all of us to varying degrees. In the following Essays I speak of the mystical experience of others, of my own enlightening experience, and the insights gained thereby; and of the near impossibility of adequately communicating that knowledge to others.

8.

AGNOSTICISM EXAMINED

This word, "agnosticism", was originally coined by Thomas H. Huxley in 1869. It appears that Huxley's intellectual path to the recognition that "agnosticism" (*not-knowing*) best described his own position regarding the existence of God was inextricably bound up with his lifelong exposure to the doctrines of Christianity, and the doubts concerning those doctrines which were surfacing in the intellectual culture prevalent in his day. It was a time when the atheistic writings of Auguste Comte (1798-1857), the anti-metaphysical writings of Immanuel Kant (1724-1804), and the evolutionary theories of Charles Darwin (1809-1882) were all current and culturally influential. In fact, after Darwin's *On The Origin of Species* was published in 1859, Huxley quickly adopted Darwin's views, and became a most effective public spokesman for Darwin's Theory. It is very probable, therefore, that the embers of Huxley's agnosticism were fanned by the naturalist's suggestion that the origin and evolution of life was simply a process of 'natural selection' requiring no assistance from God.

In the context of what he wrote at the time, it is clear that Huxley meant to imply by the word, *agnosticism*, two different meanings: (1) that he didn't know whether or not there was a God; and (2) that he believed that it was impossible to attain such knowledge. Now, this is a much more skeptical position than it first appears, for, to say that it is not possible for man to know God is tantamount to saying that there is no God. For if there *is* a God, surely He is able to reveal Himself to man, and so the possibility of knowing Him is a real possibility; but, if there is no God, it is certainly not possible to know Him. So, clearly, by asserting that it is not possible to know God, one is asserting a judgment that there is no God, and therefore contradicting one's claim to being merely ignorant of the truth, and uncommitted one way or the other.

If one's agnosticism is simply the personal acknowledgment that "I don't know if there's a God or not," then it seems to me to be a perfectly legitimate and honorable position upon which to stand if that is an accurate statement of one's subjective uncertainty. However, when the meaning of *agnosticism* is extended to include the assertion that "such knowledge is humanly unattainable", it is clear to me that what is being asserted is no longer merely a statement of personal ignorance regarding the existence or non-existence of God, but a positive belief in a universal limit on human knowledge. To say 'I don't know, and nobody else does either', assumes a little more than anyone may confidently assume. It may or may not be correct in all instances. The first part of the sentence is an honest statement of fact about one's own subjective experience, and we are on fairly solid ground; but the second part of that sentence is an inference concerning the subjective experience of others, and there we find ourselves on very slippery ground.

The point I wish to make is that, like Huxley, we all tend to assume, without any basis for that asumption, that our own experience is the norm for all. 'I haven't known God; therefore no one has known God – well, maybe Jesus did, but that doesn't count!' I would like to suggest to everyone who may read these words that, yes, there have been exceptions to that rule that no one has known God; perhaps more exceptions than you might imagine. Jesus *did* see God. He was twenty-eight, and he deliberately went into the wilderness outside of the city to seek within himself the vision of God. And he obtained that vision. Another, by the name of Siddhartha, in ancient India, was twenty-eight when, having left his home in the city, he sat beneath a tree in the jungle and looked within himself for the eternal One. He also obtained that vision. A Roman, known as Plotinus, at the age of twenty-eight, sought to know within himself that same underlying Source; and he obtained that vision as well. A young man named Stanley, at the age of twenty-eight, went into the mountain forests of Santa Cruz, California to look within himself to find God. And he also obtained that vision. Who is that omniscient person who says that each of these young men came away empty, without obtaining the vision he sought?

Those with a knowledge of such things can tell you of many, many more throughout history, from various lands, right up to the present time, who searched for God within themselves, and declared themselves victorious in their search. Many others there are who were struck – out of the blue, so to speak – with the discovery of God without even seeking Him. And if you ask them all if this is a certain knowledge and not just a supposition or misty fleeting thought, they will swear to you that the knowledge is true and real, more certain than any merely sensual knowledge could be. And yet if, relying on your own clear sense of unknowing, you state unequivocally that they are mistaken, that such knowledge is impossible of attainment, well then, you must be right; and I congratulate you on your cleverness. You know what you know, and what you don't know can't be known. That's certain enough, isn't it?

The Vatican Council of the Catholic Church says that "God can, by the natural light of human reason, be known with certainty from the works of creation" (*Const. De Fide, II, De Rev.*). No doubt, this 'rational' knowledge suffices for many. If it is not sufficient for you, and you require

The Divine Universe

God's direct 'revelation', then follow in the footsteps of those who claim to have bathed in the light of that revelation, the illumined seers, the few. And if you find that prospect a little too extreme for you, well, you can always just sit back and declare that the knowledge of God is impossible of attainment, that you are, in fact, an agnostic. That will surely give you a satisfying sense of reasonableness, and mark you among your fellows as a most humble and reasonable man.

9.

HOW DO WE KNOW?

One of the recurring problems of philosophy involves the question 'What is knowledge—and how do we define it?' The various answers to this question constitute the branch of philosophy known as *epistemology*, a subject that has been much discussed and argued throughout history. It was a question frequently discussed among the early Greek philosophers, such as Plato and his teacher, Socrates, who held that the highest and most worthy kind of knowledge was the knowledge of the Divine Ground, the *Noumenon*. However, over time, the idea that such a knowledge was at all possible of attainment fell out of favor. Also, the subjective (undemonstrable) nature of such knowledge made it suspiciously untrustworthy to some minds, and in time it became popular to regard only that knowledge whose evidence was sensory as valid, because it was experientially apparent and demonstrable. Sensory knowledge, i.e., the confirmation of sight, hearing, smell, touch, etc., came to be regarded, therefore, as the only acceptable criteria of "knowledge". Knowledge obtained in this way was considered to be *empirical* knowledge. Webster's New World Dictionary defines "empiricism" as "(1) relying or based solely on experiment and observation [the empirical method] rather than theory; (2) relying or based on practical experience without reference to scientific principles." In these sentences the sensory nature of "experiment," "observation", and "practical experience" is implied and understood, although contemporary science relies more on instrumentation as a sensory extension to verify experimental results.

But such a limited definition of "knowledge" leaves little room for a subjective, non-sensory knowing, such as the self-evident knowledge *I am*; i.e., the knowledge of consciousness. It also does not account for the knowledge of the thoughts and images existing only in the psyche; nor does it account for what we call "spiritual" knowledge. After all, we use the words "I know" to represent an inner certainty based on the various kinds of evidence to which we have access.; and this may pertain not only to sensory phenomena perceived as objects, but also to mentally perceived as well as spiritually perceived noumena. "Knowing", we must admit, is ultimately a subjective and intangible thing, difficult to put one's finger on. All forms of knowledge—even that we refer to as "empirical knowledge"—exist only as ephemeral conditions within the subjective field of awareness of each individual. And all these kinds of knowledge—empirical, mental, and spiritual—are informed by the kinds of evidence appropriate to each.

Evidence, in the scientific, empirical sense, consists of sense data. This refers, usually, to that data which reaches us through the faculty of vision via the *physical eye*. Even when there is mathematical proof of a scientific theory, empirical proof demands the confirmation of visual or instrumental measurement. But there are other kinds of *knowledge*, and other faculties of vision which provide the evidence for those other kinds of knowledge. There is not only the physical vision, but also psychic vision and spiritual vision, corresponding to the physical (phenomenal) field of experience, the mental (psychological) field of experience, and the spiritual (noumenal) field of experience; and the instruments of these various kinds of vision are the physical eye, the psychic eye, and the eye of Spirit.

The contemporary author and Buddhist mystic, Ken Wilber, has borrowed these three categories from the Medieval Christian Saint Boneventure, and has written extensively about these different faculties and instruments of vision in several of his books, namely, *Eye To Eye, The Marriage of Sense And Soul,* and *The Eye of Spirit.* In these books, he points out that, for several centuries, Western society has accepted only the empirical knowledge of science, consisting of the study of phenomena which can be verified by the physical eye; and has failed to recognize the existence of the other two kinds of knowledge. He stresses that, without acknowledging these different ways of seeing and *knowing*, we are limited to a very incomplete and woefully deficient theory of knowledge; with them, we are able to account for the entire spectrum of knowable experience—physical, mental, *and* spiritual.

Now, while the criterion of empirical proof is limited to sensory experience—specifically, *the physical eye*, the criterion for the 'proof' of dreams, imaginations, and other mental phenomena is, not physical vision, but *psychic vision*. This 'vision' does not occur through a faculty of sense, but through a faculty of the mind, or psyche, inherent in all, and is subjectively accessible by everyone: it is frequently referred to as *the mind's eye*, a term we use to represent that psychic instrument of vision whereby we "see" the images which we willfully project upon our inner 'screen' as "imaginations". It is by this inner projection that we are able to create in an instant whatever images we desire to enjoy within our own private screenings. In dreams, also, we see subconsciously produced images that our dreaming selves believe to be real while they are being presented to us. Some also claim to experience images in the waking state that are clairvoyant or prescient, or projected from other human sources, living or dead. All these kinds of visual experience are 'seen' in the mind's eye. (*Conceptual, verbal thought* seems to be audial, however, rather than visual; bringing up the likelihood of the existence of a physical, mental, and spiritual "ear" as well.)

The third kind of vision, Spiritual vision, is not obtained by means of the physical eyes, or any of the other senses, nor through the imaginative or psychic faculty referred to as "the mind's eye"; but rather through a yet subtler faculty arising only in the higher reaches of contemplative concentration, which is usually referred to as *the spiritual eye* or "the eye of contemplation". The spiritual eye "sees", but without the physical sense of eyesight or the deliberate projection of mindsight. The individual's interior awareness is lifted beyond his/her mental field of awareness, as well as beyond the awareness of worldly perceptions, as that awareness is transformed into a timeless, spaceless awareness of identity with the limitless and eternal Consciousness from which the universe emanates. In a uniting of the separative individual consciousness with the absolute and eternal Consciousness, one's awareness transcends, not only the senses and the imaginative faculty, but the sense of self, the egocentric identity, as well, relieving the individual of the sense of a separate identity, as he becomes aware of the all-inclusive One. The individual knows this eternal Consciousness as his own, since there is no separation by which he can perceive this Consciousness as other.

It is this unitive experience that we must consider the only valid knowledge, proof and confirmation of the existence of God or Spirit. No other kind of vision is appropriate to this kind of knowledge. It has long been accepted as axiomatic that reason, in the form of philosophy or metaphysics (psychic knowledge), is powerless to provide a credible proof of the existence of God, since it is limited to mental vision only, and the Spirit cannot be seen by the psychic eye; but God has been "seen" repeatedly in the unitive vision by the eye of Spirit. It should therefore be widely understood and accepted that the only self-evident knowledge and acceptable proof of God is the direct unitive vision. For those who fail of that, there is belief or faith.

The unitive 'vision' bears with it a unique kind of clarity, possessing an unmistakable and indelible stamp of truth, that does not accompany the mere physical or psychic kinds of vision. If it fails of the established standard for "knowing", then it must itself replace that standard, for it is the very essence of knowing. However, except in the case of highly gifted spiritual 'masters', it is a knowledge that is non-transmittable, and therefore undemonstrable. It may be verbally described, but that scarcely constitutes the actual direct 'knowledge' itself. It is a knowledge obtainable only via the eye of Spirit. It should immediately be added that the unitive vision must never be regarded by its recipient as a matter of pride, for it is not a deed to which the individual may lay claim. Such experience is brought about entirely by the One in whom the individual exists. The individual is not meritorious in experiencing the unitive vision; rather, he is illumined despite himself. He is drawn as if by a magnet to the experience by the power of the greater Self, and, as a dream-character in a dream is dissolved in the waking consciousness of the dreamer, his sense of separate selfhood (ego) is likewise dissolved in the wakeful Consciousness of the One in whom he lives and moves and has his being. The One alone has absolute being, and alone has effective revelatory power. It is that One who is seen, and it is that One who sees Himself in that unitive vision. Ultimately, no other may truly be said to exist but that One who exists absolutely and forever.

Many have experienced the unitive vision who have never sought it. It comes, at times, when least expected, during moments of introspective reflection, or when viewing a restful scene, or while feeling especially content or joyful. If the individual so illumined is fortunate, that unitive vision will take up perhaps twenty minutes of his life. But, for the rest of his life, his mind will hover about that vision, as a moth about a flame, in search of a continual clarification of the illuminative understanding obtained in that fleeting vision. It is in this way that he revisits the unitive vision, basking in the contemplation of the One who illumined his heart. There he finds the adoration, the bliss, and the sweet wisdom which that Self revealed to him, ever living and ever new. It is not just a memory, but it is a lasting presence in his life, benefiting him every moment, and shedding as well some little benefit to others whom he touches with his words. That vision is a lifelong treasure, filling his mind with a never-failing fountain of love and happiness and the brightest consolations of wisdom. Though to the world such a person may appear empty and alone, he possesses within himself the fullness of the universe, and his solitude is the blissful aloneness of the only One.

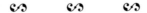

10.

ENLIGHTENMENT AND GRACE

In his several books, the highly respected psychologist and philosopher, Ken Wilber, offers a detailed and well thought-out conceptual framework for understanding and talking about the fundamental levels of experience: spiritual, mental and physical, corresponding to mystical, psychological, and scientific (empiric) knowledge. I wish to acknowledge Mr. Wilber's superior analytical vision and the very helpful framework of understanding which he has provided. However—and there is always a "however"—we are individuals with decidedly different personal proclivities, sensibilities, and styles, and there are bound to arise a number of areas in which we see things slightly differently.

Wilber stresses in most of his writings that the perennial vision of the mystics is of a hierarchic (or, more acurately, *holarchic*) reality, which he refers to as 'the great Chain of Being', in which each whole is nested in its higher (subtler) level of reality, with the non-dual One at its summit. The one absolute Source, being unqualified and indivisible, is the Ground and hierarchical whole (holon) of all that follows from It; but It is also the evolutionary Goal toward which all conscious beings are drawn. Thus, there is an involution of Spirit that can be described, in its simplest form, as a descent from Spirit to mind to matter. And evolution is the process in reverse. We may regard this Spiritual paradigm as "the perennial philosophy".

What, then, are the implications of the perennial philosophy (as derived from spiritual vision) for empirical science? In other words, how can we reconcile the data derived from the subtle vision of the mystics with the data of scientific theory? Is relativity compatible with the vision of the mystics? Is quantum theory? Or does Heisenberg's Uncertainty Principle and the stochastic nature of quantum data preclude any possible comparison of 'scientific' theory with the data perceived in the mystical vision? Does the mystic's vision of the universe as a Thought-construct offer any useful insights into an explanation of the four forces of nature? Does it offer any insights into the nature and behavior of wave/particles? At present, it yet remains to be seen whether or not all these theoretical 'phenomena' can be reconciled with the mystic's vision.

One of the difficulties in reconciling observable phenomena and physical laws with the subtle universe described by the mystics is the fact that there are subtle "layers" of reality within the mystic's universe which are not observable or verifiable in any way—such as the *soul* and *karma*; i.e., invisible causal factors (hidden variables) which are considered to largely determine spacetime realities, but which themselves are unobservable, and therefore undemonstrable, and unverifiable. The empirical 'laws' of physics bear no recognizable relationship to the 'laws' of psychology—if there are any such laws; why then should the laws of physics bear any relationship to the laws of *Spirit*, which is a yet subtler holarchic level? Of course, they are all interrelated; the physical is nested in the mental, and the mental is nested in the Spiritual. The Spiritual world is the greater holon in which these other levels reside. So, it would seem that, ultimately, both the psychic (mentally perceived) and the physical (sensorially perceived) worlds must be directly relatable to and consistent with the data obtained in the Spiritual vision.

Ultimately, science and gnosis must coincide! In fact, it should be obvious that empirical science can never succeed in formulating a 'complete' model of reality until it takes into account the mental and spiritual aspects of reality as a whole. Even if it comes up with a 'Theory of Everything', as it frequently does, it means by that term 'a Theory of Everything Phenomenal'. And even if that Theory were to be empirically demonstrated to be accurate and consistent, it would then have to recognize that only a small part of the larger reality had been explained, and that an explanation must now be found for the existence of those phenomena and noumena existing in the higher (subtler) levels of psychic (mental) and Spiritual reality. For, the perceivable, phenomenal, universe is simply an epiphenomenon of the two subtler realms. Each is related to the other, holarchically, and none may be regarded as an isolated field of enquiry.

It's a top-down universe, each level dependent on its holarchic precedent; and, ideally, knowledge of this universe must also be top-down. Ideally, we must first know the Source, the Cause; then the products, the effects, will become correctly known and understood. It is true, as Mr. Wilber points out, that the knowledge of the Source takes place, not on the sensual or mental levels, but on the spiritual level; and not with physical or psychic vision, but with spiritual vision. But there must be a means to correlate (on the mental level) the data obtained in these apparently disparate realms. If we start at the bottom, with the empirical data of the phenomenal universe, and attempt to infer from it the higher holons of reality, the mental and the spiritual, we have no consistent and reliable clues by which to infer those higher realities. In other words, when we ignore or deny the Source, as many scientific materialists and materialistic scientists presently do, it is little wonder that the theories of empirical science often go so incredibly far astray of the truth of reality as perceived in the Spiritual vision. Our understanding of the manifest, phenomenal universe requires a context; and that context can only be found at the summit of the holarchic reality; i.e., in the Spiritual vision. With that as the starting point, one may then comprehend the phenomenal reality; without it, one is left with no contextual framework at all. And that epitomizes the state of confusion and alienation prevalent in the exclusively empirical view of the world currently embraced by contemporary science.

However, in the past and in the present, Spiritual knowledge—*direct* Spiritual knowledge—has been, and it appears that it will continue indefinitely to remain, a kind of knowledge obtained by the very, very few. It is no doubt the 'highest' knowledge possible, providing a direct apperception of the summit of the holarchy of knowledge, and doubtless represents the eventual summit of human evolution; but the *universal* human apperception of the spiritual reality is a culmination that remains a long, long way off. For now, the revelation of that direct unitive knowledge occurs only in isolated instances, and the recipients of that knowledge are nearly as culturally isolated as was Jesus and Philo Judaeus two thousand years ago; though there is admittedly a noticeable increase in the *philosophical* (mental) interest in mysticism in today's world.

As I stated earlier, there are some areas in which Mr. Wilber and I differ slightly. It is evident that his concern over the current emphasis in our society on the validity of empirical (scientific) knowledge to the complete exclusion of other areas of knowledge, and the failure of the representatives of empirical knowledge to acknowledge the validity of the transcendent knowledge of the mystics, is a concern that we both share. However, one of the differences between our views that comes to mind involves Mr. Wilber's notion that there is a tried and true 'scientific' methodology for producing mystical experience, or 'the vision of God'; namely, the practice of meditation or contemplation. In several of his books, Mr. Wilber makes the pertinent point that, just as an empirical scientist must perform an experiment in accordance with the prescribed conditions of the experiment (the "injunction"), a spiritual experimenter, likewise, must conform to the injunction setting out the conditions of the spiritual experiment, namely, the practice of meditation or contemplation, in order to obtain the experiential results; i.e., spiritual vision.

This analogy to scientific empirical experimentation provides a great corrective to those who might say, 'I have not experienced spiritual vision'; whereupon one may counter, 'Well, have you conformed to the conditions prescribed for obtaining spiritual vision? Have you practiced meditation?' And if they cannot answer, 'Yes' to that question, then they simply have not fulfilled the conditions necessary for obtaining the desired results. This is all well and good. But I would like to suggest that the acquisition of spiritual knowledge through spiritual vision is not *entirely* analogous to the acquisition of empirical knowledge; and I would like to point out, in the interest of clarification, the ways in which they are different, so as to alleviate any misunderstandings resulting from the omission of this information elsewhere.

What is wrong with the logic of the following statement? 'All those who have experienced the unitive vision have done so while in a state of meditative or contemplative awareness; therefore, if you practice meditation or contemplation, you will experience the unitive vision.'? It should be clear to everyone that the concluding portion of this statement is a *non sequitur*. It just does not follow logically. It seems evident to me that if spiritual knowledge were simply a matter of fulfilling the conditions necessary for its occurrence, such as establishing a disciplined program of meditation, the world would already be filled with enlightened souls. But it is not simply a matter of fulfilling conditions, comparable to the requirement for obtaining empirical results. I do indeed wish it were true, Mr. Wilber; but it is not—and that's been the fly in the ointment all along. Is spiritual knowledge really an objective obtainable, and "perfectly repeatable", by anyone simply by setting up the prescribed conditions? Because I have 'known' God, the absolute Ground of all reality, does that mean that, by following my 'methodology' you also will come to know God? In other words, can anyone obtain the same spiritual knowledge as another simply by following certain conditional injunctions, or is the acquisition of spiritual knowledge much more dependent upon a 'Higher Will' than upon our own determined will and actions?

Einstein knew the mathematical proof of the constancy of the speed of light, and the variability of the measurement of time relative to an observer; but can you also independently arrive at that knowledge? And the answer, it seems to me, is "Only if, by the grace of God, you have the same innate inclination and the same degree of mathematical training to investigate these matters, *and* you follow the necessary injunctions for obtaining that knowledge. Otherwise, you must take it on faith that it is known." What about Beethoven? He knew how to create extraordinary music; does that mean that you also know how to do that? Same answer: 'Only if, by the grace of God, you have the same innate inclination and the same degree of musical training, *and* you follow the necessary injunctions for obtaining that knowledge.' Darwin knew that various species were related, but evolved differently over time; but can you also discover previously unknown laws of nature? Only if, by the grace of God, you have the same innate inclination and the same degree of scientific training, *and* you follow the necessary injunctions for obtaining that knowledge. This same line of reasoning may be applied to Jesus, the Buddha, Plotinus, and all other seers of the 'spiritual' reality. You may know what they knew only if you have the same innate inclination and the same degree of spiritual training, *and* you follow the necessary injunctions for obtaining that knowledge, *and* it is God's will.

It should be clear to everyone that we are not all equally capable of 'knowing' what has been known by uniquely extraordinary beings. Everything depends on our innate inclination and our specialized training, and of course the grace of God. By "innate inclination" I mean the soul-driven proclivities and talents constituting the karmic tendencies possessed by each soul. These 'innate inclinations' are wholly dependent upon the evolutionary development of our souls; which are, in turn, dependent upon, not only our own wills, but the grace of God. And so, we must acknowledge that the subtle spiritual knowledge that has been obtained by a few extraordinary men and women is not necessarily available to everyone; there must be a congruence of inclination, training, and God's grace, along with the practice of meditation or contemplation. The assertion by many spiritual teachers that the realization of God, the knowledge of the Source and Goal of all existence, is available to everyone simply by following certain precepts and injunctions, is not at all an accurate assessment. One's soul, which is itself a product of God's grace, must contain an innate inclination to the acquisition of such knowledge, must follow a regimen of training, and, by the grace of God, be placed in the most timely and appropriate cosmic circumstances to receive such knowledge. Then, and only then, will it be able to 'know' God. Is God-realization available to everyone? I think not. I think that, not only spiritual knowledge, but each kind of knowledge: sensory, mental, or spiritual, is available only to those whom God has made fit for it; it is misleadingly inaccurate to say that such knowledge is available to everyone.

The injunctions given by Jesus, "Seek and ye shall find," "Knock, and the door shall be opened to you", has inspired many followers to seek and to knock, and yet we must wonder, how many of those millions of followers were enlightened with the unitive vision of God? I can think of only a handful of Christians who seem to have obtained this vision over the past twenty centuries. The injunctions given by the Buddha, "Meditation brings wisdom; therefore, choose the path of wisdom", has drawn many to meditation; and some have become illumined—but only a small percent. My point is that there is no guaranteed means or methodology for obtaining the unitive vision. It seems to me to depend on many factors, not all of which are within the purview of one's own will. It would certainly be wonderful if one could truthfully and confidently say 'Do this, and you will experience the unitive vision', but in spiritual matters there is no direct causal relationship between voluntary acts and revelation such as there is between empirical injunctions, spelling out the conditions of the experiment, and the produced results. 'Do this, and that will result' is sound and dependable advice when we are advising "release the ball, and you will see that it falls to earth"; but not necessarily as truthfully predictive when we are advising "practice meditation, and you will become enlightened". If it was an easily reproducible experience, it is likely that enlightenment would have been widely accepted as a readily obtainable and commonly repeatable experience by now—which is certainly not the case.

It is no doubt true that one living in an environment conducive to meditation has an advantage over one who is immersed in a turbulent and disturbing environment, but we must not leap to the conclusion that all the monks in the temple, monastery or ashram are therefore enlightened. The one thing we can say for certain is that they are exposing themselves to the lifestyle and practices conducive to the unitive vision. It is not because the Buddha sat down under a Bo tree to meditate that he became enlightened; it is not because Jesus went alone into the wilderness to pray and contemplate God that he became illumined; it is not because John of the Cross gave himself to introspection and prayer within his Toledo cell that he was united with God. All of these mystical seers found themselves drawn to conditions that were amenable to that experience, but the underlying Cause was the grace of the all-governing Spirit, which called each soul from within to evolve toward the egoless reception of that non-dual revelation; in other words, it was God's singular grace which was the ultimate causative factor in that revelation. I am aware that this is an unpopular stance; but experience has taught me that the revelation of the unitive vision cannot be reduced to a causal act initiated by the individual.

Indeed, we need to ask ourselves, "Who is this 'I' who thinks it can bring about the realization of the transcendent God by its own efforts?" It is well known that only when this false and limited 'I' is vanished is the revelation of God at all possible! And by whose grace do you suppose the death of that false 'I' is accomplished? Whose love wells up in the soul and draws it to that immolation? And whose 'I' is revealed in the unitive vision as the Ground and essence of all 'I's? If you think you can bring this about by your own efforts, go right ahead. As Saint Nanak has said, eventually, 'suffering will teach you wisdom'.

The 'causes' of grace cannot be discussed, of course; because only the One is privy to the factors that go into its bestowal. I am of the opinion, however, that, in His universe, "all things move together of one accord", and that many elements must come together in the production of the revelation of the soul's higher Identity. There is a coordinated unfoldment in the manifested world of one's mental, emotional, and karmic conditions along with the conditions of the physical environment, and the positions of the planets in the cosmic environment—all under the watchful and governing eye of the Spirit—to bring about that unitive vision. In other words, man proposes, but God disposes. None may deliberately, willfully transcend and supercede His unerring Will. When it is that soul's time for enlightenment, he will be drawn from within to seek it; he will be drawn to the conducive location; he will be drawn into spiritual communion, and he will be illumined in his soul by the Light of the one Spirit. (Please see, in relation to this, my Essay entitled, "The Astrology of Enlightenment")

Innumerable saints and seers have declared their utter dependence upon God's grace in obtaining spiritual vision; here are just a couple: Saint Nanak, the *Adi* (original) *Guru* of the Sikh tradition (1469-1539 C.E.), used to say, "By God's grace alone is God to be grasped. All else is false, all else is vanity." In one of his songs, addressing God, he reiterates this conviction:

> He whom Thou makest to know Thee, he knows Thee;
> And his mouth shall forever be fulll of thy praises.
> ...Liberation and bondage depends upon Thy will;
> There is no one to gainsay it.
> Should a fool wish to, suffering will teach him wisdom. [1]

Another seer, named Dadu (1544-1603 C.E.), was also eloquently unambiguous in declaring ths truth:

> Omniscient God, it is by Thy grace alone that I have been blessed with vision of Thee.
> Thou knowest all; what can I say?
> All-knowing God, I can conceal nothing from Thee.
> I have nothing that deserves Thy grace.
> No one can reach Thee by his own efforts; Thou showest Thyself by Thine own grace.
> How could I approach Thy presence? By what means could I gain Thy favor?
> And by what powers of mind or body could I attain to Thee?
> It hath pleased Thee in thy mercy to take me under Thy wing.
> Thou alone art the Beginning and the End; Thou art the Creator of the three worlds.
> Dadu says: I am nothing and can do nothing.
> Truly, even a fool may reach Thee by Thy grace. [2]

These examples could be multiplied extensively, and I would add my own declaration to the list. However, I think one could compile a much more lengthy list of those who, having practiced meditation for many years, did *not* experience an enlightening revelation, did *not* obtain the unitive vision. So, I feel that the suggestion that enlightenment follows a cause-effect sequence that anyone may experientially prove to his or her own satisfaction simply by the practice of meditation is a useful tool for encouraging the search for enlightenment (which is no doubt its function), and it may indeed prove fruitful in specific instances. But it is also unrealistic and unreliable as an unqualified injunctive rule—unless, of course, we leave the time frame open-ended. I know of one spiritual teacher who used to tell his followers that, if they continued to practice meditation, they would be enlightened in eight, ten, or twelve lifetimes, depending on their effort. Looked at from that time frame, the guarantee appears much more plausible. The fact is, we are all, in our spiritual essence, identical with the one Spirit, the transcendent Lord of the universe; and one day all, by the grace of God, must come to know it. On that you may rely.

Notes:

1. Singh, Trilochan, et all. [eds.], *Selections From The Sacred Writings Of The Sikhs,* London, George Allen & Unwin, 1960; *Rag Asa,* pp. 57, 42 (or see Abhayananda, S., *History of Mysticism,* London, Watkins, 2002; pp. 335-344).
2. Orr, W.G., *A Sixteenth Century Indian Mystic,* London, Lutterworth Press, 1947; p. 142 (or see Abhayananda, S., *History of Mysticism,* London, Watkins, 2002; pp. 345-356).

ఌ   ఌ   ఌ

11.

MY OWN EXPERIENCE [1]

My little cabin in the redwoods was cool in the summer, but damp in the winter, as I discovered that first winter in '66. The little babbling brook swelled to a cascading Colorado river in my backyard, and I had to catch water coming down the slope from the road in little waterfalls to get clear water for drinking or cooking. Each night I sat close to the cast-iron cooking stove, with the little side door open so I could watch the dancing blue and gold flames sizzle the oak logs and turn them to glowing ash.

Day and night, during the California winter, the rain drizzled outside the window in a steady, gray, time-dissolving continuum. In the mornings, I'd prepare oatmeal and a bath by the stove; I'd pour hot water from a pitcher over my body onto the concrete floor, and then sweep it outside. The rain would stop sometimes during the day, and then I would go out and walk the once dusty logging roads through the woods and up through the meadows in the high ground. "Hari! Hari! Hari!" was my continual call.

The dark skies kept my energies subdued, and my mind indrawn. My days passed uneventfully. It was in the night that the embers of my heart began to glow keenly as I sat in the dark, watching the fire contained in the stove. A stillness—sharp-edged and intense—filled my cabin and I spoke very closely, very intimately, with the God who had drawn me there. And He would sometimes speak to me in the stillness of the night, while I wrote down His words.

Hari became my only thought, my only love. And while the days and nights became endless stretches of grayness, wetness, my mind became brighter and brighter with an intense light that displayed every wandering thought that arose as a compelling drama in bold Technicolor and Panavision; and then I would pull my mind back with "Hari!" I had realized that I could have or become whatever I settled for in my mind; and I was determined to refuse every inspiration that was not God Himself. I was steadfastly resolved to refuse all envisualizations, all mental wanderings, holding my mind in continual remembrance and longing for Hari alone.

In the evening twilight, I'd sing to Him, to the tune of *Danny Boy*:

> O Adonai, at last the day is dying;
> My heart is stilled as darkness floods the land.
> I've tried and tried, but now I'm through with tryin';

The Divine Universe

> *It's You, it's You, must take me by the hand,*
> *And lead me home where all my tears and laughter*
> *Fade into bliss on Freedom's boundless shore.*
> *And I'll be dead and gone forever after;*
> *O Adonai, just You, just You alone, forevermore.*

Or, sometimes, I'd sing this song, to the tune of *Across The Wide Missouri:*

> *O Adonai, I long to see you!*
> *All the day, my heart is achin'.*
> *O Adonai, my heart is achin';*
> *O where, O where are you?*
> *Don't leave me here forsaken.*
> *O Adonai, the day is over;*
> *Adonai, I'm tired and lonely.*
> *My tears have dried, and I'm awaitin'*
> *You; O Adonai,*
> *You know I love you only.*

Sometimes, to focus my mind on Him, to bring devotion to my sometimes dry and empty heart, I'd read from Thomas á Kempis' *Imitation Of Christ*—a version which I had pared down from the original; and this had the invariable effect of lifting my heart to love of God, and brought me, as though by sympathetic resonance, to the same simple devotion and purity of heart evidenced by that beatific monk of the 15th century. I felt so much kinship with him, so much identification with him, that I came to love his little book above all other works for its sweet effect on me.

Then, deep into the night, I'd sit in silent prayer; my wakefulness burning like a laser of intensely focused yearning, a penetrating, searching light-house of hope in the black interior of the cabin, as I witnessed the play of the flickering flames dying out in the stove's interior. On one such night, filled with Divine love, the understanding came to me that it was just this Love that was drawing me to Itself within me. It was this Love that was the Soul of my soul, calling me to live in Its constant light. I lit a candle; a song was being written in my notebook, and I was understanding very clearly, very vividly, just what it was that I loved, what it was that I was pledging my life to:

> *Thou art Love, and I shall follow all Thy ways.*
> *I shall have no care, for Love cares only to love.*

Spiritual Vision

I shall have no fear, for Love is fearless;
Nor shall I frighten any, for Love comes sweetly and meek.
I shall keep no violence within me,
Neither in thought nor in deed, for Love comes peacefully.
I shall bear no shield or sword,
For the defense of Love is love.
I shall seek Thee in the eyes of men,
For love seeks Thee always.
I shall keep silence before Thine enemies,
And lift to them Thy countenance,
For all are powerless before Thee.
I shall keep Thee in my heart with precious care,
Lest Thy light be extinguished by the winds;
For without Thy light, I am in darkness.
I shall go free in the world with Thee—
Free of all bondage to anything but Thee—
For Thou art my God, the sole Father of my being,
The sweet breath of Love that lives in my heart;
And I shall follow Thee, and live with Thee,
And lean on Thee till the end of my days.

<u>November 18, 1966:</u>

This was the night I was to experience God. This was the night I learned who I am eternally. All day long the rain had been dripping outside my cabin window. And now the silent night hovered around me. I sat motionless, watching the dying coals in the stove. "Hari!" my mind called in the wakeful silence of my interior. During the whole day, I had felt my piteous plight so sorrowfully, so maddeningly; "Dear Lord, all I want is to die in Thee," I cried within myself. "I have nothing, no desire, no pleasure in this life—but in Thee. Won't you come and take this worthless scrap, this feeble worm of a soul, back into Thyself?"

"O Father," I cried, "listen to my prayer! I am Thine alone. Do come and take me into Thy heart. I have no other goal, but Thee and Thee alone."

Then I became very quiet. I sat emptied, but very awake, listening to God's silence. I balanced gingerly, quakingly, on the still clarity of nothingness. I became aware that I was scarcely breathing. My breath was very shallow, nearly imperceptible—close to the balance point, where it would become non-existent. And my eyes peered into the darkness with a wide-eyed intensity that amazed me. I knew my pupils must be very large. I felt on the brink of a meeting with absolute clearness of mind. I hovered there, waiting. And then, from somewhere in me, from a place deeper than I even knew existed, a prayer came forth that, I sensed, must have been installed in my heart at the moment of my soul-birth in the mind of God: "Dear God, let me be one with Thee, not that I might glory in Thy love, but that I might speak out in Thy praise and to Thy glory for the benefit of all Thy children."

It was then, in that very moment, that the veil fell away. Something in me changed. Suddenly I knew; I experienced infinite Unity. And I thought, "Of course; it's been me all the time! Who else could I possibly be!" I lit a candle, and by the light of the flickering flame, while seated at the card table in my little cabin, I transmitted to paper what I was experiencing in eternity. Here is the "Song" that was written during that experience (the commentaries in parentheses which follow each verse were added much later):

O my God, even this body is Thine own!

(Suddenly I knew that this entity which I call my body was God's own, was not separate from God, but was part of the continuous ocean of Consciousness; and I exclaimed in my heart, "O my God, even this body is Thine own!" There was no longer any me distinct from that one Consciousness; for that illusion was now dispelled.)

Though I call to Thee and seek Thee amidst chaos,
Even I who seemed an unclean pitcher amidst Thy waters—
Even I am Thine own.

(Heretofore, I had called to God in the chaos of a multitude of thoughts, a multitude of voices and motions of mind—the very chaos of hell. And in my calling, I was as though standing apart from God; I felt myself to be an unclean pitcher immersed in the ocean of God, dividing the waters within from the water without. Though God was in me and God was without, there had still remained this illusion of 'me'. But now the idea of a separating 'ego' was gone. And I was aware that I—this whole conglomerate of body, mind, consciousness, which I call "I"—am none else but that One, and belong to that One, besides whom there is nothing.)

Does a wave cease to be of the ocean?

(A wave is only a form that arises out of the ocean and is nothing but ocean. In the same way, my form was as a wave of pure Consciousness, of pure God. How had I imagined it to be something else? And yet it was that very ignorance that had previously prevented me from seeing the truth.)

Do the mountains and the gulfs cease to be of the earth?

(Mountains and valleys in relation to the earth, like waves in relation to the ocean, seem to have an independent existence, an independent identity; yet they are only irregularities, diverse forms, of the earth itself.)

Or does a pebble cease to be stone?

(A pebble is, of course, nothing but stone—just as I now realized in growing clarity that I was none else but the one 'stuff' of Existence. Even though I seemed to be a unique entity separate from the rest of the universe, I was really a piece of the universal Reality, as a pebble is really a piece of stone.)

How can I escape Thee?
Thou art even That which thinks of escape!

(Thought too is a wave on the ocean of God. The thought of separation—can that be anything but God? The very tiniest motion of the mind is like the leaping of the waves on the ocean of Consciousness, and the fear of leaping clear of the ocean is a vain one for the wave. That which thinks of separation is that very Consciousness from which there can never ever be any separation. That One contains everything within It. So, what else could I, the thinker, be?)

Even now, I speak the word, "Thou," and create duality.

(Here, now, as I write, as I think of God and speak to Him as "Thou," I am creating a duality between myself and God where no duality exists in truth. It is the creation of the mind. Having habituated itself to separation, the mind creates an "I" and a "Thou," and thus experiences duality.)

I love, and create hatred.

(Just as for every peak there's a valley, so the thought of love that arises in the mind has, as its valley, as its opposite, hatred. The impulse of the one creates the other, as the creation of a north pole automatically creates a south pole, or as "beauty" necessitates "ugliness," or as "up" brings along with it "down," or as "ahead" gives birth to "behind." The nature of the mind is such that it creates a world of duality where only the One actually is.)

I am in peace, and am fashioning chaos.

(The very nature of God's phenomenal creation is also dual; His cosmic creation alternates from dormant to dynamic, while He, Himself, remains forever unchanging. In the same way, while our consciousness remains unmoved, the mind is in constant alternation. For example, when it is stilled, it is like a spring compressed, representing potential dynamic release. The mind's peace, therefore, is itself the very mother of its activity.)

Standing on the peak, I necessitate the depths.

(Just as the peak of the wave necessitates the trough of the wave [since you can't have one without the other], wakefulness necessitates sleep, good necessitates its opposite. Exultation in joy is paid for with despair; they are an inseparable pair.)

But now, weeping and laughing are gone;
Night is become day.

(But now I am experiencing the transcendent "stillness" of the One, where this alternation, this duality, of which creation is made, is no more. It is a clear awareness that all opposites are derived from the same ONE, and are therefore dissolved. Laughing and its opposite, weeping, are the peak and the trough which have become leveled in the stillness of the calmed ocean, the rippleless surface of the waters of Consciousness. Night and day have no meaning here: All is eternity.)

Music and silence are heard as one.

(Sound, silence—both are contained in the eternal Consciousness which cannot be called silent, which cannot be called sound; It produces all

sounds, yet, as their source, It is silence. Both are united in the One of which they consist.)

My ears are all the universe.

(There is only Me. Even the listening is Me.)

All motion has ceased;
Everything continues.

(The activity of the universe does not exist for Me, yet everything is still in motion as before. It is only that I am beyond both motion and non-motion. For I am the Whole; all motion is contained in Me, yet I Myself am unmoving.)

Life and death no longer stand apart.

(From where I am, the life and death of individual beings is less than a dream—so swiftly generations rise and fall, rise and fall! Whole eons of creation pass like a dream in an instant. Where then are life and death? How do they differ? They too are but an artificial duality that is resolved in the One timeless Self.)

No I, no Thou;
No now, or then.

(There is no longer a reference "I" that refers to a separate individual entity; there is no longer anything separate to refer to as "Thou." This one knowing Consciousness which is I is all that exists or ever existed. Likewise, there is no "now" or "then", for time pertains only to the dream and has no meaning here beyond all manifestation.)

Unless I move, there is no stillness.

(Stillness, too, is but a part of duality, bringing into existence motion. Motion and stillness, the ever-recurring change, are the dream constituents in the dream of duality! Stillness without motion cannot be. Where I am, neither of these exists.)

Nothing to lament, nothing to vanquish;

(Lament? In the pure sky of infinity, who is there to lament? What is there to doubt? Where there is no other, but only this One, what error or obstacle could there be? What is there to stand in the way of infinity? What is there other than Me?)

**Nothing to pride oneself on—
All is accomplished in an instant.**

(Pride belongs only to man, that tiny doll, that figment of imagination who, engrossed in the challenge of conflict with other men, prides himself on his petty accomplishments. Here, whole universes are created in an instant and destroyed, and everything that is accomplished is accomplished by the One. Where, then, is pride?)

All may now be told without effort.

(Here am I, with a view to the Eternal, and my hand writing in the world of creation, in the world of men. What a wonderful opportunity to tell all to eager humanity! Everything is known without the least effort. Let me tell it, let me share it, let me reveal it!)

Where is there a question?

(But see! Where everything is very simply and obviously Myself, what question could there be? Here, the possibility of a question cannot arise. Who could imagine a more humorous situation?)

Where is the temple?

(What about explaining the secrets of the soul, and how it is encased in that temple of God called 'the body?' That secret does not exist; for, when all is seen and experienced as one Being, where is that which may be regarded as the receptacle, the temple?)

**Which the Imperishable?
Which the abode?**

(Which may I call the imperishable God, the Eternal? And which may I call the vessel in which God exists and lives? Consciousness does not

perish. The Energy of which this body consists does not perish. All is eternal; there is no differentiation here.)

I am the pulse of the turtle;
I am the clanging bells of joy.

(I am everywhere! I am life! I am the very heartbeat of even the lowliest of creatures. It is I who surge in the heart as joy, as surging joy like the ecstatic abandonment of clanging bells.)

I bring the dust of blindness;
I am the fire of song.

(I am the cause of man's ignorance of Me, yet it is I who leap in his breast as the exultation of song.)

I am in the clouds and I am in the gritty soil;
In pools of clear water my image is found.

(I am that billowing beauty in the sky; I play in all these forms! And the gritty soil which produces the verdure of the earth—I am that soil, that black dirt. I am every tiny pebble of grit, cool and moist. And when, as man, I lean over the water, I discover My image, and see Myself shining in My own eyes.)

I am the dust on the feet of the wretched,
The toothless beggars of every land.

(I live in the dust that covers the calloused feet of those thin, ragged holy men who grin happily at you as you pass them by.)

I have given sweets that decay to those who crave them;
I have given my wealth unto the poor and lonely.

(Each of my manifestations, according to their understanding, receives whatever they wish of the transitory pleasures of the world; but the wealth of My peace, My freedom, My joy, I give to those who seek no other wealth, who seek no other joy, but Me.)

My hands are open—nothing is concealed.

(I have displayed all My wealth; according to his evolution, his wisdom, each chooses what he will have in this life.)

All things move together of one accord;
Assent is given throughout the universe to every falling grain.

(All is one concerted whole; everything works together, down to the tiniest detail, in the flower-like unfoldment of this world. All is the doing of the One.)

The Sun stirs the waters of My heart,
And the vapor of My love flies to the four corners of the world.

(Like a thousand-rayed sunburst of joy, My love showers forth as the universe of stars and planets and men. And then, this day of manifestation gives way to the night of dissolution ...)

The Moon stills Me, and the cold darkness is My bed.

(And the universe withdraws into My utter darkness of stillness and rest.)

I have but breathed, and everything is rearranged
And set in order once again.

(The expansion and contraction of this entire universe is merely an outbreath and an in-breath; a mere sigh.)

A million worlds begin and end in every breath,

(And, flung out into the endless reaches of infinity, worlds upon worlds evolve, enact their tumultuous dramas, and then withdraw from the stage once more. This cycle repeats itself again and again; the universe explodes from a single mass, expands as gas, and elements form. Eventually they become living organisms, which evolve into intelligent creatures, culminating in man. And one by one each learns the secret that puts an end to their game. And again, the stars reach the fullness of their course; again everything is drawn back to its source...)

And, in this breathing, all things are sustained.

∞

Spiritual Vision

After this, I collapsed in bed, exhausted by the sheer strain of holding my mind on so keen an edge. When I awoke, it was morning. Immediately, I recalled the experience of the night before, and arose. I went outside to the sunlight, dazed and disoriented. I bent, and took up a handful of gravel, letting it slip slowly through my fingers. "I am in this?" I asked dumbfoundedly.

I felt as though I had been thrust back into a dream from which I had no power to awaken. My only thought was to return to that state I had known the night before. I rushed up the twisted road and scrambled up the hill to the cliff on top of the world, above the forest and ocean, where I had often conversed with God; and I sat there, out of breath, praying, with tears running down my cheeks, for Him to take me back into Himself. Before long, a chill blanket of gray fog, which had risen up from the ocean below, swept over me, engulfing me in a misty cloud. And after a few moments, I reluctantly went back, down the mountain.

Notes:

1. *This Essay is excerpted from a more detailed account of my spiritual journey in my book, The Supreme Self, Olympia, Wash., Atma Books, 1998; subsequently published by O Books, London, 2005.*

12.

THE GIFT OF SPIRITUAL VISION

For the *bhakta,* the soul in the throes of love for God, there arises a love-longing for the union with God. And prior to the dawning of that unitive experience, there is much singing and prayers, and copious tears. But then, at the inception of the experience of revelation, there is an end to the emotion, and the soul falls into a calm that is also intensely awake. The pupils of her eyes become extraordinarily open wide, and her breathing slows and subsides to a very shallow rise and fall, as though it were approaching the balance point, where breathing would be entirely stilled. All relationship of soul to God is vanished, and there is only the fine awareness focused upon its own incredible clarity, its own being; and then the prayer that bursts forth from the finally naked and surrendered soul: "O God, let me be one with Thee—not that I might glory in Thy love, but that I might speak out in Thy praise and to Thy glory for the benefit of all Thy children". And then comes the sudden awakening, as though from a dream. And you are seeing with the eyes of the eternal One, who is the Self you have always been. You, who have been crying for His embrace; you, who were awaiting the arrival of the King; you, *yourself,* are the only Existence, the Lord, the Father; and all along you had been living in an illusory separation from yourself, in a dream-world of your own making. *Even now, I speak the word, 'Thou', and create duality.* There is no one else, and never had been; you are the omnipresent Mind—you! The personification you had adopted was but a fantasy; and now you see the truth. You live eternally, showering forth this huge universal display. *I am the pulse of the turtle; I am the clanging bells of joy. I bring the dust of blindness; I am the fire of song. I am in the clouds and in the gritty soil; in pools of clear water my image is found. ...I have but breathed, and everything is rearranged and set in order once again. A million worlds begin and end in every breath, and in this breathing, all things are sustained.*

The prayer that precipitated this vision was the prayer of a soul, still caught in the illusion of separation; yet the desire to praise God was God's desire speaking through the soul, and in this life she has no other purpose but to honor that prayerful desire. It permeates this soul, and constitutes her task in this life, her only joy. It may be that she was given no mandate from God to teach. But it was she who asked to be united with Him in order that she might speak out truly in His praise and to His glory. And that desire sprang from the deepest place in that soul, a soul which is itself fountained forth from God. And so that desire was truly His desire in her. His granting of that desire for intimacy constituted His mandate. When she looks at the lives and missions of others before her, like the Egyptian author of the Hermetic teachings, like the Buddha, Jesus, Plotinus, etc., she associates strongly with the sense of mission each possessed, having been graciously lifted up to intimacy with God, and filled with the desire to praise Him. What a singular grace, and what responsibility it confers! Yet, despite the gift of this advantageous vision, all were mere mortals, with the limitations that implies. All had to endure the earthly life of bodily provision, sickness and death; and all had to endure the doubt and malevolence of the community of other men and women. Yet still they communicated their vision as best they could. Their lifelong desire to see and to give expression to the truth of God is God's enduring gift to us, His wondrous, thrillingly beautiful gift of overwhelming joy to all of us.

And once the larger, subtler, eternal reality is known, the soul, returned to awareness of this world, can scarcely see the phenomenal reality in the same way as before. During the visionary experience of the Eternal, she is identical with the Eternal, and blissfully content to remain in that state. However, that state wanes and gives way to the return in consciousness to this temporal and phenomenal reality. This is truly an unwelcome eviction. Having known the bliss of her eternal Self, she is at first greatly shocked and dismayed at finding herself back in this little world of separable images in time and space. But after her initial dismay, she reflects on her current state, and quickly realizes that she is still the eternal Self, and that the world to which she has returned consists solely of the bright Energy breathed forth from her Divinely transcendent Self. She recognizes that now she is in a dream-movie, but it is the dream-movie of God, who is indeed her very Self; and even this body in which she moves about is woven of that Divine fabric.

She realizes that, even in this projected image which God puts forth, she remains enveloped in His blissful Being, and realizes that she could never be anything but safe at home in Him. That is the great gift of Spiritual vision: that now she sees this transient world of images as suffused with ethereal light, and splayed with dazzling beauty. Joyful contentment fills the air she breathes, and adoration fills her heart. This is the translation of divine Spiritual vision into the world of phenomenal awareness. This is the carryover from the transcendent vision to the sensory vision here on earth. She carries over from that higher realm no intellectual understanding of how a photon operates as both a particle and a wave, or how the force of gravity interacts with the moving earth. Let physicists puzzle over these dusty details; she is content to see the beauteous God in evidence all about her and within her. To abandon that untold treasure of joy to pick and peck amidst the crumbs of reason's paltry scrapings would be but the conduct of a fool. You can have it, you mathematicians and quantum mechanicians! You biochemists and cosmo-theoreticians! It's all been settled and displayed to her utmost satisfaction: Beauty beauty beauty everywhere, and the wine of intoxicating nectar in her cup! What needs she more?

And yet, having seen O so clearly that all the beings who exist on earth are truly embodiments of the one Divine Self, the desire to share this wondrous knowledge remains an insistent urge deep within her soul. But also she is aware that each soul follows an evolutionary path unique to itself, and is able to comprehend the presence of God only in its own time, and only by the gracious gift of God. And so her words have relevance now and in the future only to those whose eyes are already opened, to those on whom God's grace has already shone. Then rejoice with her, all ye fortunate souls! And be merciful to those whose temporary blindness is also His gift. He will lift that blindness in His time, and release all from the darkness in which they now live. He will open to their eyes, as He did to hers, the light and warmth, the wonder and delight, the beauty and the joy, of His immeasurable life-giving Love. Praise God!

13.

WE WHO HAVE BEEN BLESSED

If we reason clearly and correctly, we must come to the acknowledgement of our utter dependence on God's blessings. We have no power, no intelligence, no sweetness, and no illumination of our own; all that we know as ours is the gracious gift of God. For that reason, we cannot claim to have earned spiritual wisdom or vision by some worthiness of our own making. Whatever comes to us does so of His power and His grace. And so, though I would gladly offer instruction and advice in the endeavor to assist others in joining their soul to God, I am too clearly cognizant of the fact that He alone can bring each soul to His embrace; and that He alone, whose Light illumines all, can peel away the blinders of the illusory ego, and show Himself as the true and everlasting Self of all.

So, what am I to say to those who ask the way to God? 'Follow the noblest that's in you; that will lead you home to Him. Revere silence, solitude, reflection, and deep thought. Read the lives and words of those who found their way to His door, and thus purify your heart. Above all, converse with Him; He'll guide you from within and lead you every inch of the way. Rejoice often in His great love in looking after you and all good souls. He is the inner life, the inner heart, of you; and He seeks only what is your highest joy and light. Turn your face to Him, your mind to Him, your heart to Him; and nothing else at all needs to be done. When it is your time to know your eternal identity, the whole universe, including the stars in the heavens, will conspire to bring about your awakening. Do not fear; no one will be forgotten or left behind.'

14.

HE HEARS

In my book, *The Supreme Self*, I told of my retreat to a small cabin in the mountains of Santa Cruz, of my hermit life, and my subsequent mystical experience; and when I first approached my publisher with this book, I was put on the defensive by his question, "Do you think everyone should do the same as you have done?" In reply, I asked him, "Do you think everyone should be a book publisher?" The point to be made is that everyone has their own proclivities along with their own unique task in this life. That I may want to be a concert pianist does not necessarily imply that I feel that this is what everyone should do. Look at the life of Jesus, for example: while he is certainly exemplary in many respects, it does not follow that everyone should attempt to replicate the events of his life in their own. If we believe that there is a natural evolution at work, by which each individual learns in each lifetime what he or she needs to learn in order to progress toward the ultimate Good, then we must allow that there are different unique life paths for each individual, and that each will follow the path that naturally appeals to his or her self-revealed nature.

There are some few who are drawn to the religious life: some to a life of service, some to a contemplative life, or to a mixture of the two. It is not a field so lucrative that it attracts competitors desirous of material gain. Rather, it is a path upon which one enters in order to follow an inner yearning for the knowledge and service of God. It is a yearning inspired by a grace known inwardly in the soul. There are some who, following this inner calling, obtain a further grace: that of vision. In a moment of prayer or deep contemplation, the mind becomes focused above its normal plateau, and finds itself staring into the normally unfathomable depths of its own consciousness wherein lies the fundamental source of all that exists. In that vision, one's own nature and the nature of all existence is revealed, as one becomes lifted in consciousness to a union with the eternal Mind that we call 'God'. From that vantage, there is no longer a soul and a God, for the two are then one wakeful seeing, one eternal Being.

After some time, the mind, no longer able to retain that height, no longer able to remain fixed in that intensely one-pointed focus, sinks by its own weight away from that supremely attractive delight. Yet it retains the afterglow of that divine visitation, awed and inebriated by the infusion of knowledge and joy that revealed itself in him. It is a soul once more, limited to a single body, cast back into an alien environment, still longing to return to that timeless and unbounded country. There, he is what he has always been; there, he is the true and unfigured Self of which he is now but an imaged copy. And now, having been thrust once again back into the throng of selves in this busy world, would he not be urged from within to tell of what he had seen for the benefit of all whose source and destination he now knows full well? Surely, all would wish to know what had been revealed of that hidden source!

Ah, though speak he might, in this shadowland very few are able to hear him. The pride of life spreads over all, concealing in its deadening roar the sound of the true seers' words, and hiding in its cloaking mirage the knowledge of the single father of us all. The people go on, unheeding, uninterested in what our visionary has to say. But that is how it's always been; perhaps that's how it will always be. For it is clear that God is hidden by His own design, and it is He who makes Himself known. The game goes on; the others too must find the breadcrumbs scattered here and there, and follow clues to come at last into His vestibule. All is indeed well. Was this not finely shown in the clarity of his vision? (*Nothing to lament, nothing to vanquish, nothing to pride oneself on; all is accomplished in an instant.*) The all-inclusive One brings all to fruition in His own time, by His mysterious yet merciful ways. Nevermind that no one hears; *He* hears, and governs all. No need to fret or fear. He holds us all, and brings us, one by one, along our way to home in Him.

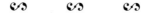

III.

THE PERENNIAL PHILOSOPHY

"The perennial philosophy" is a phrase coined by Gottfried Wilhelm Leibniz (1646-1716), and popularized by Aldous Huxley in his book of the same name. It is a multi-faceted philosophy to be sure, and one not meticulously defined. But its primary tenet is that direct religious experience is the common core and unifying factor in all religious traditions. Such experience reveals the identity of the experiencer with the eternal and ultimate reality, a realization frequently expressed in philosophical terms as 'Nonduality'. Here, then, in this third grouping of Essays, philosophy attempts to systematize the insights gained through inner vision, and can only make generalizations. In the end, words fail us, and we must merely live our philosophy. If we retain the words, it is only because we long to share our understanding.

15.

KAPILA'S VISION

All mystical philosophy worth its name is a product of vision, spiritual vision. One of the earliest philosophical expressions of spiritual vision came from a man who lived in India somewhere between the 7th to 9th century B.C.E. His name was Kapila, and the philosophy by which he gave his vision expression is known as *Sankhya* (wisdom). From that time and for many centuries thereafter *Sankhya* was the adopted spiritual philosophy of all India. It forms the philosophical foundation of the sacred scripture known as the *Bhagavad Gita* or "Song of God", (*ca.* 500 B.C.E.), whereby most Westerners are introduced to Kapila's philosophy.[1] There, as well as in the *Srimad Bhagavata Purana,* it provides the basis for the teachings expounded by the avatar, Krishna. Kapilavastu, the town in which Siddhartha, the Buddha, was born in 586 B.C.E. was named for Kapila, testifying to the sage's widespread fame at that time.

Kapila's philosophy is unquestionably Nondualistic, yet it introduces a semblance, an appearance, of duality in the two terms: *Purusha* and *Prakrti*. To interpret these two terms in the simplest manner, let us render them as God (or Spirit) and His Power of universal manifestation. These two, having separate labels, appear to be two separate realities; but, clearly they are one and undivided. *Purusha* represents the transcendent aspect of God, the Absolute Ground, the Godhead, the eternally undivided and undifferentiated Consciousness; and *Prakrti* represents the creative Power by which He casts forth the Energy of which the entire universe consists, and by which He is immanent in all phenomenal reality. For those unillumined by "the vision of God", however, there appears to be only the Energy, whose manifestation we call 'matter'. They do not see the Source and Author of this Energy, nor are they able to intuit that Source; therefore, they regard the Energy and Its manifestations as the sole existent.

Let me back up for a moment, and provide for you some deep background on this subject: All mystical philosophy must deal with the question of 'How does a God described as absolute and unqualified also act to create the phenomenal universe?' And the answer given by every Nondualistic philosophy based on mystical vision is that the one Reality possesses two distinguishable aspects. Each of the different authors of the different mystical philosophies appearing at different times and places throughout history have come up with their own names for these two "aspects"; but they are referred to in one way or another as 'the supreme Absolute' and 'Its creative Power'. The Supreme is beyond all descriptive categories, though It is sometimes referred to as *Sat-chit-ananda*, "Existence, Consciousness and Bliss". The creative Power of the Supreme is not a separate thing, but is merely Its own Power of effluent production. The distinction between these two is similar to the distinction drawn between an individual mind's consciousness and its image-producing faculty. Though one is the source of the other, they are essentially one and the same.

In the mystical vision, the highest Truth, the ultimate Reality, is seen to be pure Consciousness, with no qualifying characteristics. It is the Highest, beyond which nothing is. And yet It produces, emanates, radiates, or projects a manifest Cosmos from Itself. Such productive ability is analogous to the ability of an individual human mind to project thoughts and images upon its own interior screen. And just as the human consciousness is the source and unmanifest substratum of all thought production, the One, the absolute Consciousness, is the Source and Substratum of all that follows upon It. If the One were content to remain merely inactive, no universe would be produced; but It utilizes Its inherent Power to breathe forth a dramatic outpouring of Energy by whch this universe is created.

Thus, the mystics from the earliest times and the most widely diverse cultures have described their visions in a similar and comparable manner; having recognized these two somewhat dissimilar aspects of the eternal One, they have perennially labeled these two with separate and distinct names. For example, at least a thousand years before Kapila, some mystic living among the inhabitants of the city of Mohenjo-Daro in northern India called these two by the names "Shiva" and "Shakti". Several hundred years later, a mystic from among the invading Aryans called them "Brahman" and "Maya". In China, during the sixth century B.C.E., a mystic by the name of Lao Tze called these two aspects of the Eternal by the names "Tao" and "Teh". In Greece, Heraclitus (540-480 B.C.E.) called them "Zeus" and "Logos"; the mystics of a struggling band of nomads in Judaea called them "Yahveh" and "Chokmah".

The occasional mystics of each culture with its own religious language recognized these two aspects of the Divine by giving each of them names reflective of their own culture and language. The tenth century Arab, al-Hallaj, called them *Haqq* and *Khalq*. Meister Eckhart, a thirteenth century German, called them *Gottheit* (Godhead) and *Gott* (God, the Creator). We, today, would call them the transcendent *Consciousness*, and Its creative Power or *Energy*. But make no mistake, the names are only names; whether you call that eternal One by the name of *Brahman, Allah, Shiva, Yahveh, Nirvana, Tao* or *Zeus*, It is the same; and if you call the creative Energy that shines forth

The Perennial Philosophy

from Him by the name of *Shakti, Chokmah, Samsara, Logos, Maya,* or any other name, It is still the same. As stated long ago in the *Vedas,* "Reality is one; sages call It by various names."

Now, back to Kapila: Kapila called these two aspects of the One, *Purusha* and *Prakrti*. As noted, these two had previously been termed *Shiva* and *Shakti* by the indigenous population within his own culture; and the Vedic (Aryan) culture which entered India around 2000 B.C.E. called them *Brahman* and *Maya,* or *Vishnu* and *Shri (Lakshmi)*. Popular culture objectified them in figurative form as a male sovereign and his female consort; and so Kapila was not introducing anything new. *Purusha* (God) is the transcendent and invisible Consciousness, the absolute Ground from which *Prakrti* (His Power) emanates, producing the universe of form. Therefore, all that we experience in this world is a projection of the Power of *Purusha* (the Divine Consciousness), i.e., His *Prakrti* (Energy/Matter). [2]

Kapila further envisualizes *Prakrti* (God's Energy) as consisting of three different strands, or *gunas:* a positive (active) energy (*rajas*), a negative (inert) energy (*tamas*), and a neutral (balanced) energy (*sattva*). As there has since been found no empirical evidence for their existence, it is easy to view the notion of the *gunas* as merely a simplistic and naïve pre-scientific hypothesis, meant to account for the human mental and physical tendencies of activity, torpidity, and serenity, which otherwise seem to fluctuate within a person unexplainably. And while the *gunas* have been interpreted in many ways, to me they suggest the positive, negative, and neutral charges which, according to modern Western physics, characterize the three modes in which energy manifests in its most elementary form, and which combine in an infinite variety of assemblages to make up the elemental wave/particles of matter.

But, putting aside consideration of whether or not the concept of the *gunas* is a literal or a figurative representation of the constituency of the universal Energy, we must see that Kapila's essential understanding of a nondual Reality, producing from Itself a Mind-born phenomenal universe, is in accord with all spiritual (mystical) visions and is the archetype of all mystical theologies. All those who have "seen" or experienced their Selfhood as identical with the Supreme, regardless of religious affiliation or linguistic orientation, have reported a common understanding. The *subject* of such a vision perceives that his/her consciousness and permanent identity is none other than the one transcendent Source (*Purusha*) of all, and that his/her body as well as all other material phenomena that constitute the *objects* of perception consists of the projected Energy (*Prakrti*) of that Source. Therefore, the spiritual Self of man is said to partake of *Purusha,* the eternal Self, while his material aspect is *Prakrti* (*Purusha's* image-projecting Power).

The Divine Universe

Thus, in this and in the expression of all other authentic Nondual mystical philosophies, the apparent duality arising between the conscious mind and the body is resolved in the one ultimately indivisible Reality.

When this issue is confronted by Western secular philosophy, however, it results in a great muddle. Take Descartes for example: He asserts that there are two substances: mind and body; but that, while they have the ability to effect and influence one another, they remain separate and distinct entities. Descartes was a great thinker, but he was not a mystic; while he possessed a strong conviction of the existence of God, he never spoke of having attained anything like spiritual vision. He was a rationalist, worshipful of the intellect; and so it never dawned upon him that his vaunted "I", the self/mind/soul of man, labelled by his detractors as "the ghost in the machine" of the body, is the Divine Consciousness, traceable to the very Godhead Itself; and that the "created" material phenomena (including the body) is a projection of that universal Mind, existing within the transcendent Mind, as a dream-figure exists within the mind of a man. But this realization lies beyond the perception of the intellect, and is, alas, inaccessible to the conceptualizing mental faculties of man, being accessible only to the revelation of spiritual vision.

However, Western philosophy, since the late Medieval period, has rejected the reliability of Divine revelation, and recognized only rational investigation as the method appropriate to the pursuit of certain knowledge. Later, with Kant's demolition of the validity of all epistemological methods save empirical proofs, even the Rationalist search for certain metaphysical knowledge was abandoned; and then, even later, with Heisenberg's insights, the empirical approach also was seen to be impotent to produce certainty. Philosophy today, echoing the pronouncements of empirical science, declares that human knowledge, in all its forms, is doomed by its very nature to fall short of absolute certainty. No wonder there is in evidence today such a deep perplexity and anxiety among the thoughtful, and such a sinking and degradation of the directionless populace in a mindless stupor of sensuality, chaos, and violence.

Still, within the different religious traditions, there exists and has always existed a small cadre of spiritual devotees possessing certainty in the awareness of the Divine inner Self. Invoking Him by many names, they worship in their hearts the one transcendent Spirit who manifests this cosmic array by His own Power. They call It by various names, such as *Theos* and *Logos*, *Tao* and *Teh*, *The One* and *Nous*, *God* and *Mother Nature*, and many other pairs of names; yet, recalling Kapila, we may recognize these two as *Purusha* and *Prakrti*, the eternal Consciousness and His universal Energy, and know them as the complementary and inseparable aspects of our own divine Self and all that is.

Notes:

1. For the finest and most perfect rendition of the *Bhagavad Gita*, see Paramahansa Yogananda, *God Talks Wish Arjuna: The Bhagavad Gita, Royal Science of God-Realization*, Los Angeles, Self-Realization Fellowship, 1995.

2. All that is accomplished in this world is accomplished by *Prakrti*, the Divine Energy; the transcendent aspect, *Purusha*, remains in Its eternal quietude and perfection. All the cities, all the hubub, the wars, the great poetry, and this book as well—all are the work of *Prakrti*. We, who are the instruments of this activity, imagine that they are our personal works, and say "I did this"; but there is no "I" other than *Purusha* manifesting as *Prakriti*. *Prakrti* is doing everything. And, we mustn't forget: *Prakrti* is the Conscious Energy of God (*Purusha*); all is His doing. There is no other.

☙ ☙ ☙

16.

THE PHILOSOPHY OF NONDUALISM

Spiritual Nondualism [1] is the philosophy, substantiated by the vision of all the mystics who ever lived, which holds that the soul or self of man is identical with the transcendent Spirit that is God. *Spirit* is a word which is not very much in favor these days, as it has been sorely misused in the past; but it is a necessary word, as there is a need for a non-sectarian verbal symbol to represent the essence of our reality as experienced by the visionaries of the subtle. *Spirit* represents the Divine in both Its transcendent and Its immanent aspects. As the transcendent Godhead, Spirit is incognizable and inconceivable by the mind of man. It is neither mind nor matter, but a subtler reality that is eternal, omniscient, and omnipresent; and which experiences Itself as pure consciousness and bliss. The transcendent Godhead manifests Itself as the Spirit immanent in man and the universe, appearing as the consciousness of self in man, as well as the Divine Energy that goes to make up all the 'material' forms of this universe. There is but one Being, one Spirit, who constitutes both the Eternal and the temporal reality. Just as the Sun's rays are identical with the Sun, all that emanates from the Spirit is identical with the Spirit. In other words, our true identity is Spirit, and nothing other than Spirit. This identity may be known directly, by the grace of God, when the soul is drawn to that unitive interior vision.

Nondualism means that there is no difference between the source and the manifestation, no difference between the essence of one thing and another, no difference between you and God. No difference. The Sanskrit word used by the Upanishadic sages to designate this philosophy is *advaita*, which is made up of *a*, not, and *dvaita*, two; meaning literally, "not two". *Spiritual Nondualism* is the philosophy that you are essentially identical with the one Spirit—that One who has been called *Brahman, Shiva, Allah, Yahveh, Hari, Adonai, Karim, God*, and innumerable other names throughout history. The Upanishadic seers, recognizing this identity in their visions, have called that one Spirit the *atman;* the Self, as it is realized to be the one Self of all. Now, this vision belongs to no one religion. Though the philosophy of Nondualism was originally elucidated in the Upanishads, today, it is universally recognized by the mystics (the seers) of every religious tradition as the *Perennial Philosophy*, because it arises again and again throughout history as the one recurring view of mystical philosophers and seers from widely divergent cultures and traditions. It has been and continues to be repeatedly verified through direct experience by all those who have made the ascent in consciousness to the supreme Self.

Here are a few snippets of quotations from some well known seers representing that perennial philosophy:

He who is beyond all predicates appears as the relative universe; He appears as all sentient and insentient beings.
– Rig Veda

Even by the mind this truth is to be learned: there are not many, but only One.
– Katha Upanishad

It is not what is thought that we should wish to know; we should know the thinker. "He is my Self!" This one should know. "He is my Self!" This one should know.
– Kaushitaki Upanishad

The pure man sees the One as one and the many as one. So long as he sees the Unity, he is God; when he sees the distinctions, he is man.
– Chuang Tze

He who knows that he is, himself, the Lord of all, and is ever the same in all, immortal though experiencing the field of mortality, he knows the truth of existence.
– Bhagavad Gita

The best of men choose to know the One above all else; it is the famous "Eternal" within mortal man.
– Heraclitus

What, then, is the heart of the highest truth, the core of knowledge, the wisdom supreme? It is "I am the Self, the formless One; by my very nature, I am pervading all. That one God who shines within everything, who is formless like the cloudless sky, is the pure, stainless Self of all. Without any doubt, that is who I am."
– Dattatreya

The Divine Universe

God is high above place and time... He is contained by nothing, but transcends all. But, though transcending what He has made, nonetheless, He fills the universe with Himself.
— Philo Judaeus

There is one Divine Reality, divided as higher and lower; generating Itself, nourishing Itself, seeking Itself, finding Itself. ...It is both Mother and Father, a Unity, being the Root of the entire circle of existence.
— Simon Magus

I and the Father are one.
— Jesus

The one Divine Mind, in Its mentation, thinks Itself; the object of Its thought is nothing external; Thinker and thought are one, unchangeably the same.
— Plotinus

All that is He contains within Himself like thoughts: the world, Himself, the All. In the All there is nothing which is not God. Adore this teaching, my child, and hold it sacred.
— Hermes Trismegistus

The Reality is One; though, owing to illusions It appears to be multiple names and forms, attributes and changes, It always remains unchanged. [It is] like gold, which while remaining one, is formed into various ornaments. You are that One, that Brahman. Meditate on that in your mind.
— Shankara, *Vivekachudamani*

Just as the light of the Sun and the Sun are not absolutely different, ... so also the soul and the supreme Self are not different.
— Shankara, *Vedanta Sutras*

The entire universe is truly the Self. There exists nothing at all other than the Self. The enlightened person sees everything in the world as his own Self, just as one views earthenware jars and pots as nothing but clay.
— Shankara, *Atma Bodha*

The Perennial Philosophy

When the mystery of the oneness of the soul and the Divine is revealed to you, you will understand that you are not other than God. ...For when you know yourself, your sense of a limited identity vanishes, and you know that you and God are one and the same.

– Ibn Arabi

My Me is God, nor do I recognize any other Me except my God Himself.

– Catherine of Genoa

As the soul becomes more pure and bare and poor, and possesses less of created things, and is emptied of all things that are not God, it receives God more purely, and is more completely in Him; and it truly becomes one with God, and it looks into God and God into it, face to face as it were; two images transformed into one. ...Some simple people think that they will see God as if He were standing there and they here. It is not so. God and I, we are one. ...By the living God, it is true that there is no distinction!

– Meister Eckhart

When I clutched at His skirt, I found His hand in my sleeve. ...I am the one I love; He whom I love is I.

– Iraqi

He to whom all things are One, and who draws all things into One, and sets all things in One, and desires but One, may soon be stable in heart and be fully pacified in God.

– Thomas á Kempis

The world in which we live is a play of Chiti Shakti, the self-luminous universal Consciousness. For a man who sees this, the world is nothing but a play of God's energy. ...Chiti plays in the external world and yet stays ever the same. ... In Her unity, She is supreme Shiva, supreme Consciousness, absolutely alone. In this mode, She is called the transcendent supreme Shiva, the "formless, attributeless Absolute" of the Vedantins. She has two aspects: the supremely pure transcendent aspect, which is above the world, and the immanent aspect, in which, by Her own desire, She becomes the universe within Her own being.

– Swami Muktananda

All of the above statements were written by mystics who had directly experienced the oneness of which they spoke. Nondualism is above all a philosophy based on direct experience; without that direct vision, the philosophy itself is of little value. That direct experience confers on its recipient the blissful knowledge of the Divine Self in its fullness. It is referred to as "Self-realization", "God-realization", "yoga", or simply "enlightenment". It occurs only rarely, by the grace of the Spirit, and usually in association with a regimen of introspective meditation or contemplation. And, because that direct experience is limited to a select few, the philosophy of Nondualism has never been accepted as a major cultural worldview by the greater populace, but continues to live on perennially as the spiritual philosophy of an elite spiritual intelligentsia. The main opposition to its broader aceptance comes from the blind exponents of materialism and the unillumined partisans of various sectarian religious faiths.

Notes:
1. There is also another kind of Nondualism: *Material Nondualism*. The philosophy of *Material Nondualism* is one with which all of us are familiar, since we are bombarded with its tenets every time we pick up a book on contemporary "Science". Material Nondualists believe that there is but one existent that makes up the source and substance of this universe and all that exists within it: and that one thing is *matter*. How it is possible to hold such a view is completely beyond me. However, there are some who do manage to hold this view by ignoring the question of what produced the singularity from which they claim all matter was born, and by ignoring the absence of an adequate answer to the question of how consciousness arose from matter. But no matter! We are here focusing on another kind of Nondualism: *Spiritual Nondualism* or *Idealist Nondualism*. Spiritual Nondualism is the conviction that the soul or self and God are not essentially different. It is a philosophy of the Nondualism of the *Spirit*.

☙   ☙   ☙

17.

PERFECT NONDUALISM

Part One

The philosophy of Nondualism was beautifully expressed in the Upanishads, written by some nameless sages perhaps a thousand years before the Current Era. Buddhist seers of later centuries wrote their own scriptural texts explaining an identical vision. Then, the great medieval revitalizer of Nondual philosophy, Shankaracharya (7th to 9th centuries C.E.), following in the Upanishadic tradition, set forth in very explicit terms the nature of the One without a second. In explaining the *apparent* duality between God (*Brahman*) and the world (*jagat*), he referred to the world as a product of the creative Power of God—His *shakti*, or *maya*; and asserted that the phenomenal world produced by *maya* was *mithya*, "illusory" or "unreal". The phenomenal universe, said Shankaracharya, is an *adhyasa*, a "superimposition", upon Brahman. Let me try to explain what he meant by this:

For the mystic who has experienced in himself the clear revelation of the nature of Reality, the world appears to be a superimposition upon the absolute Consciousness, very much the way a thought, dream, or mental image is superimposed upon our conscious awareness when it is produced in our mind. Consider: an image is projected from the conscious mind, is made of nothing but mind, and at the same time is superimposed upon that conscious mind. The thought-image is not the same thing as the mind; and yet, who would say that they are different? They are like the ocean and its waves, or like the Sun and it rays. They are different, and yet they are the same. One is the substratum, and the other is a transient phenomenal manifestation, superimposed upon that substratum. Thus, Shankara's use of the word "superimposition" to describe the relationship between Brahman and the world is not an unreasonable way of expressing in language this paradoxical duality-in-Unity. However, his terminology was regarded as unfortunate by many, as it seemed to imply a *real* duality between God and the world. If there is something superimposed, some reasoned, it must be something *other* than Brahman.

Shankara, in his many writings, frequently differentiated between Brahman, the eternal Self, and Maya's product, the world, simply in order to guide the earnest seeker away from attention to the transient appearance (the phenomenal world), and toward the eternal Reality (the Self). He never intended to imply, however, that the transient appearance was anything but Brahman. Here, let him explain in his own words:

> Brahman is the Reality, the one Existence. Because of the ignorance of our human minds, the universe seems to be composed of diverse forms; but it is Brahman alone. ...Apart from Brahman, the universe does not exist. There is nothing beside Him. It has no separate existence, apart from its Ground.

And again:

> The universe is truly Brahman, for that which is superimposed has no separate existence from its substratum. Whatever a deluded person perceives through mistake is Brahman and Brahman alone. The silver imagined in mother-of-pearl is really mother-of-pearl. The name, "universe", is superimposed on Brahman; what we call "the universe" is [really] nothing but Brahman. [1]

Shankara never intended to imply by the use of his word, "superimposition", that there was something other than Brahman superimposed on Brahman. But, unfortunately, that is what arises in the minds of some when they hear this word, "superimposition". Some objected vigorously to his terminology. For example, a 13th century Maharashtran saint, by the name of Jnaneshvar, objected to the notion of superimposition as an implication of duality, and attempted to clarify the doctrine of Nonduality in the following passage from his book, *Amritanubhav*, "The Nectar of Mystical Experience":

> When it is always only the one pure Consciousness seeing itself, why postulate the necessity of a superimposition? Does one superimpose the sparkle on a jewel? Does gold need to superimpose shininess on itself? [2] A lamp that is lit does not need the superimposition of light; it is resplendent with light. Likewise, the one pure Consciousness is resplendent with radiance. Therefore, without obligation to anything else, He easily perceives Himself. [3]

> ... Whatever form appears, appears because of Him. There is nothing here but the Self. It is the gold itself which shines in the form of a necklace or a coin; they, themselves, are nothing but gold. In the current of the river or the waves of the sea, there is nothing but water. Similarly, in the universe, nothing exists or is brought

The Perennial Philosophy

into existence that is other than the Self. Whether appearing as the seen, or perceiving as the seer, nothing else exists besides the Self. [4]

Perhaps it is impossible to adequately express in words the differentiation between the eternal Consciousness and Its creative Energy without making it appear that they are two separate things. This would seem to be the case, since every time one mystic gives expression to his vision, another mystic takes exception to the way it is described, and tries his own hand at it, only to have another mystic come along somewhere down the line who takes issue with *his* terminology. In any case, Shankara's writings gave rise to many misunderstandings, and to clear up some of these misunderstandings of terminology, some mystics living in the northern state of Kashmir, including a seer named Vasugupta, devised their own interpretation of the philosophy of Nondualism, which they called *Kashmir Shaivism*.

Legend tells that Vasugupta had a dream in which Lord Shiva told him the whereabouts of a large rock on which Shiva himself had inscribed some teachings in the form of brief aphorisms regarding the nature of God, the soul, and the universe. The inscriptions were copied from the rock by Vasugupta and later became known as the *Shiva Sutras*. Thus, like many other religious traditions, Kashmir Shaivism claims Divine revelation as its source. Such revealed scriptures are called *agamas* by those who embrace this tradition. Other *agamas*, besides the *Shiva Sutras*, are the *Malini-vijaya*, the *Vijnana-bhairava*, and the *Rudra-yamala*. There are also some subsidiary scriptures which explain the *agamas*; these are called *spandas*, or *spanda-karikas*, which formulate doctrine. Then there are the philosophical works which attempt to present the teachings in a logical and ordered form; these are the *Pratyabijna shastras*. Some of these are *Shiva-drshti* by Somananda (ca. 875-925 C.E.), *Ishvara-pratyabijna* and *Shivastotravali* by Utpaladeva (ca. 900-950 C.E.), and *Pratyabijna-hridayam* by Kshemaraj. The philosophy expressed in these scriptures of Kashmir Shaivism also came to be known as *Pratyabijna Darshana*, "The Philosophy of Recognition"; and also as *Purna Advaita*, or "Perfect Nondualism".

The ultimate Reality, according to Kashmir Shaivism, is *Paramashiva*, "the Supreme Shiva". This is, of course, synonymous with *Parabrahman*, "The Supreme Brahman", of Vedanta. Indeed, in all cases, there is no difference whatever between the vision of Vedanta and that of Kashmir Shaivism, except for the differences in terminology. For example, the Advaita Vedanta of Shankara holds that Brahman "projects" the world by His creative Power (*Maya*), and Kashmir Shaivism says that Paramashiva "appears" as the world through His creative Power (*Shakti*). Shankara says the universe is a "superimposition" upon Brahman; Kashmir Shaivism says the universe is simply Paramashiva apppearing as form. There is not the slightest difference between them except for their terminology. It is commonly found in this world that isolated groups of people with a common religious tradition tend to regard their own way of saying things to be more correct than the way some other people of another tradition may say it. The reality is that language, by its very nature, is imprecise; and it is the sage who knows the Truth by experienceing It directly who really knows the Truth.

The sages of Kashmir Shaivism say that Paramashiva is the one Reality; all is taking place within Him. But He remains unchanged and unmoved by all this multiplicity and apparent change. He is the transcendent Totality, and so He remains the same, no matter what. To Him, there is only the pure sky of Consciousness and Bliss. He remains awake to His oneness always, while the "creation" comes and goes. It is breathed out by Him and breathed in again, in an ever-recurring cycle. It is manifested, and then re-absorbed back into Him. This emanation is called *abhasa*, a "shining forth". Then, when it is withdrawn again, that is called *pralaya*. The complete cycle is a *kalpa*—which amounts to 4 billion, 320 million years of Earth-time.

According to the sages of Kashmir Shaivism, a *kalpa* begins with a *spanda* (what in more recent times is regarded as the impetus to "the Big Bang"). *Spanda* is the first movement of will, the initial flutter or throb of movement in the Divine Will, or *Shakti*. As for the question, "Why does He create at all?" the answer given by the Kashmir Shaivites is the same as that given by the Vedantists: "It is simply His nature to do so." It is His innate nature to breathe forth the universe of multiplicity; and yet, at the same time, it is asserted that He manifests the universe of His own free will, as a play, or sport (*lila*). In fact, the very first Sutra of the *Pratyabijna-hridayam* says that "The absolute Consciousness, of Its own free will, is the cause of the manifestation of the universe."

The Pratyabijna philosophers say that, from *spanda*, then comes the bifurcation into *aham* and *idam*, subject and object. These two aspects of the One are also spoken of as *prakasha* and *vimarsha*. *Prakasha* is the conscious light, the witness-Consciousness, the "subject" aspect of Paramashiva. *Vimarsha* is Its power of self-manifestation; i.e., the "object" aspect of Paramashiva. Thus, inherent in the process of manifestation is this Self-division of Paramashiva into conscious subject and phenomenal object; from this initial polarity, all other dualities, including manifold souls, come into being. And, according to the Kashmir Shaivite philosophy, while there is never anything but Paramashiva, the souls thus created by this Self-division experience a limitation of their originally unlimited powers. As stated in the *Pratyabijna-hridayam* of Kshemaraj, "Consciousness Itself, descending from Its universal state, becomes the limited consciousness of man, through the process of contraction. Then, because of this contraction, the universal Consciousness becomes an ordinary human being, subject to limitations."

The truth, of course, is that the Lord, the one Supreme Consciousness, is never subject to limitations. He lives in absolute freedom. He is all-pervading and all-knowing. By His Power, He can do whatever He likes. And so, in order to become many and play within the (imaginary) multiplicity which is the universe, He sheds His undifferentiated state of Unity and accepts differences. Then, His various powers of will, knowledge and action appear to have shrunk, though this is not really so. This limited state is the state of ordinary people, subject to limitations, such as you and I.

When *Shakti* manifests as individual conscious entities, the one Consciousness *appears* to be bound by Its own Self-imposed limitations; Its primal powers of omniscience, perfection, everlastingness and all-pervasiveness are then experienced in a reduced condition. Although omniscient, He knows only a few things; though omnipotent, He feels helpless and acts effectively only in a small sphere. The master of perfect Bliss, He is ensnared in pleasure and pain, attachment and aversion. The eternal Being cries aloud from fear of death, regarding Himself as mortal. Pervading all space and form, He grieves because He is tied to a particular place and a particular form. This is the condition of all creatures whose *Shakti* is reduced, and who are caught in the transmigratory cycle. Again, quoting from the *Pratyabijna-hridayam*: "To be a transmigratory being, one needs only to be deluded by one's own Shakti."

It is because Shiva, the Self, has become involved in His own Shakti—that is, manifested in form, that He finds Himself in the state of "an ordinary being, subject to limitations." But, we must see, it is His sport to do so. Without such an "involution", there could be no evolution. The evolution, or unwinding, of a watchspring could not occur without there first being an involution of the watchspring created by the winding of the watch. A log burns, i.e., evolves into energy, only because energy, in the form of sunlight, water, and soil, has become involuted as the log of wood. Evolution is the reverse transmutation of an effect into its cause. Paramashiva, or Brahman, or Chit-Shakti, has "involved" Himself in the form of gross matter, and through the human form, must "evolve" back to Himself.

It is only in the human form that one is able to choose to take the evolutionary path back to the Source, because of the development of mind. It is the mind that is capable of development toward intelligence, concentration, meditation, and, finally absorption in pure Consciousness. This is evolution. It is also known as "Liberation", as it is the freeing of oneself from identification with the body and the activity of the mind, and thus from rebirth. Liberation, or *moksha*, is freedom from the vicious cycle of births and deaths which from the beginning of creation are whirling a soul around. In fact, life is not worthy of the name, "life", as it is really no more than a series of limitations, the very nature of which pinches the soul and makes it hanker after something real, something permanent, beyond the pale of sensual pleasures and pains, something not clouded with the gloomy, lusty, desires, which are never quenched and are never satiable. Real "life" is that for which the soul yearns with an incessant longing, though not knowing where and how it is to be obtained. Still, it feels with an inborn conviction the existence of a greater life, a greater Self, as a tangible reality. Everyone yearns for it, because life, eternal life, is the soul's very nature.

The astute student will recognize the aforementioned doctrines of Kashmir Shaivism as quite consistent with the precepts of Vedanta. The ultimate goal of the "bound" soul is the knowledge of the Self, which constitutes "liberation" from the wheel of transmigration. This is the teaching of both Vedanta and Kashmir Shaivism (and Buddhism as well), revealing once again their undeviatingly common perspective. But, it is only natural that all philosophies stemming from real "mystical" experience will find agreement in nearly all their conceptual elements. Listen, for example, to what is said in the *Ishvara Pratyabijna-vimarshini* of Abhinavagupta (ca. 950-1000 C.E.):

> The knowledge of the identity of the soul (*jiva*) and God (*Shiva*), which has been proclaimed in the scriptures, constitutes liberation; lack of this knowledge constitutes bondage.

In other words, it is ignorance of our true nature that binds us, and nothing else. In fact, it is clear that we have never been actually bound. This is brought out in the *Tripura rahasya*, attributed to Dattatreya, which states:

> Though, in reality, there is no bondage, the individual is in bondage as long as there exists the feeling of limitation in him. ... In fact, there has never been any veiling or covering anywhere in Reality. No one has ever been in bondage. Please show me where such a bondage could be. Besides these two false beliefs—that there is such a thing as bondage, and that there is such a thing as mind—there is no bondage for anyone anywhere.

Both Vedanta and Kashmir Shaivism recognize the possibility of *jivan-mukta*, liberation from the wheel of transmigration while still living in the body. However, it is not merely the mystical experience of Unity which constitutes this self-liberation; one must also assimilate the knowledge thus acquired into one's everyday consciousness, and make the knowledge of the Self an ever-present awareness. Here is the statement of this ultimate liberation from the *Pratyabijna-hridayam*:

> Final realization is possible only when the complete nature of the Self is realized. Though there might be release after death, there can be no release in life unless the universal Self is grasped through the intellect. Indeed, the equanimity in the experience of worldly enjoyment and in the experience of Unity is what truly constitutes the liberation of the soul, while living. ... The individual who identifies with the Self, and regards the universe to be a sport and is always united with it, is undoubtedly liberated in this life.

And this is reiterated in the *Spanda-karika*:

> This entire universe is a sport of universal Consciousness. He who is constantly aware of this truth is liberated in this life, without doubt.

Part Two

Sadhana is the period of one's spiritual journey in search of the Self. And the *sadhana* of Kashmir Shaivism is the same as the *sadhana* of Vedanta: it consists of self-effort and Grace. Self-effort is in the form of learning about the Self, contemplating the knowledge gained, and meditating on the Self. It is a self-effort toward Consciousness; but Self-realization comes of Grace. There is nothing to be done to receive it, but to be true to the Self, to give our purified hearts to the communion with God within. In this way, we prepare ourselves for Grace.

Every great spiritual teacher, including Jesus, taught that one realizes God through His Grace alone. This may be verified in the Christian scriptures; for example, when Jesus was asked by some of his disciples, "Who, then, can enter the kingdom of God? (in other words, Who can realize the Self?)," Jesus replied, "For man it is impossible; but for God all things are possible." [5] He was saying, in other words, 'Don't ask me how to know God. It can't be done by you or me or anyone! It is God Himself who makes Himself known. Only He has the power to reveal Himself.' What we can do is to open our hearts and minds, our souls, to receive the light of His Grace; and this alone is the skill, the art, if you will, that we must acquire. The giving of His gifts is entirely in His hands. If anyone can dispute this of his own experience, and has the power to experience the Self at his own whim and convenience, I have yet to hear of such a person.

The Divine Universe

The philosophers and sages of Kashmir Shaivism hold exactly this same view; furthermore, they hold that this Grace is absolutely undetermined and unconditioned. As it is stated in the *Tantraloka* of Abhinavagupta: "Divine Grace leads the individual to the path of spiritual realization. It is the only cause of Self-realization, and is independent of human effort." If it were dependent upon some conditions, it would not be absolute and independent Grace. Grace is the uncaused Cause of the soul's release. What appears at first glance to be a condition of Grace, is, in reality, a consequence of it. For example, devotion, which may seem to bring Grace, is, in fact, the result or gift of Grace. In the Kashmir Shaivite tradition, the Absolute is said to carry on the sport of self-bondage and self-release of His own free will; and the postulation of conditions or qualifications would be against that doctrine of free will. This position is made clear in the *Malini Vijaya-vartika*:

> The learned men of all times always hold that the descent of grace does not have any cause or condition, but depends entirely on the free will of the Lord.

And again in the *Paramartha Sara*:

> Throughout all these forms, it is the Lord who illumines His own nature. In reality, there is no other cause of these manifestations except His freedom, which alone gives rise to both worldly enjoyment and Self-realization.

Here, the question may arise that if Divine Grace has no regard for the merit and demerit of the recipients, does it not amount to an act of partiality on the part of God? How is it that He favors some individuals by bestowing His Grace and disfavors others by keeping it away from them? And the answer is that Grace is operative all the time for all individuals. The difference in the descent of Grace is really the differences in the receptivity of the individual souls, each of whom evolves at his own unique pace. Moreover, this problem does not have much significance in the Non-Dualistic philosophies of Vedanta and Kashmir Shaivism; because it is the Absolute Himself who appears first as bound, and then as liberated, owing to His own free will. He cannot be accused of partiality, since it is only Himself whom He favors or rejects.

As for self-effort, this is accomplished by our inherent power of will. Shakti, the Divine power of will, exists in us in a limited form. This will, which we possess, is the faculty by which a person decides upon and initiates action. Fickleness of mind flutters and weakens the will-power; and conversely, a strong desire and one-pointed longing strengthens it. But too many desires and hankerings after many objects, and aimless running about in pursuit of sense-pleasures dissipates the creative energy, the will-power. As one clear-minded sage said, "A definite purpose of action, backed by a strong will, is a sure way to success in any endeavor. Therefore, minimize your desires, make a deliberate choice, and focus the whole energy of your will-power in that particular direction, and you will never miss your goal."

The will of a person may be made to flow in two different, and opposite, directions: outwardly, toward secular worldly goals, or inwardly, toward spiritual goals. If one wishes to concentrate one's energy toward spiritual goals, then the creative energy, the will, must be diverted from its normal outward-flowing course; by closing all such outlets in the form of worldly desires, one at last attains the state of desirelessness. Then, it is possible to turn the mind inwardly to the Self, and attain spiritual knowledge.

It is the desires for worldly objectives that distract one from the attainment of spiritual objectives. But, for one who is established in the pursuit of spiritual goals, worldly gains have little charm, and the necessary duties one must perform in the world take on a spiritual significance. To such a person, every act on the worldly plane is a service to the Lord, in the fulfillment of His will, and a stepping stone for the upward progress toward spiritual enlightenment.

Therefore, when the objective, or outward, trend of the will is checked, and is given a turn in the opposite direction, the "involved" Shakti begins its evolutionary journey; and, instead of experiencing a poverty of Shakti, a person begins to expand his or her powers, and to feel greater energy, intelligence, increased abilities and an expanded sense of well-being and completeness. Turning in the direction of its source, the mind begins to sense its identity with the Self, the pure and all-perfect Consciousness of the universe. This is the beginning of the evolution from the human to the Divine.

Now, if it were an easy thing to revert the flow of the will from worldly to spiritual objectives, everyone would be able to manage it. But it is not easy. The mind is totally deluded by the amazing and wonderful appearance spread out before it; and, unaware that it is all its own projection, it reaches out eagerly for satisfaction and pleasure from the ephemeral and empty mirage. Intellectually acquired knowledge helps us to recognize the mirage for what it is—but still, old habits must be overcome. And that is not an easy task. To subdue the habits of nature, instilled by long practice and conviction, to subdue the old outgoing tendencies of the mind, requires great effort. This is known as *tapasya*.

The Divine Universe

To understand what *tapasya* is, we must understand that it is *Shakti*, the Divine Energy, which manifests as our minds and bodies and their various activities. And, frequently, we expend that Energy in thoughtless and frivolous ways, and thus remain listless and groggy through much of our lives. But, if we could learn to conserve our natural *Shakti*, then we could reap the benefits in the form of greater physical and mental energy, and a clearer awareness of the blissful Self, our eternal Identity. *Tapasya*, which literally means, "making heat," is the restraint of the outgoing tendency of the mind and senses, which conserves and heats the *Shakti*. The *Shakti*, turned inward, then begins to nourish and invigorate the brain and the whole body, expanding one's natural powers as well as one's consciousness.

Here are some of the traditional methods of *tapasya* that help to conserve and evolve the *Shakti* toward its source, *Shiva* (the Self):

(1) *Mantra repetition:* This conserves the *Shakti* by subduing the wandering mind and the *prana*, and focusing the attention on God within.

(2) *Devotional singing:* This heats the *Shakti* through emotion, and elevates the awareness toward God. It is a form of devotional meditation that brings joy and satisfaction to the heart.

(3) *Concentration of the mind:* By deep thought, attention, study, or meditation, the *Shakti* is concentrated and focused, and the mind becomes subtle and clear.

(4) *Surrender of the fruits of actions:* This relieves the mind of futile exertions, conserving the *Shakti* and retaining the steadiness of the mind.

(5) *Eating properly, moderately and regularly:* It is the *Shakti* which is the central regulator of the mind and body; it preserves the heat and cold of the body, and distributes the effects of various foods and drinks to the different parts of the body, not only through the bloodstream, but through the nerve currents as well. The choice of a proper, moderate, and regular diet is therefore of great importance.

(6) *Continence:* When the *Shakti* has been given an evolutionary turn, and begins to flow inward and upward instead of outward through the senses, there is an accumulation of heat in the region near the base of the spine. It is there the *Shakti* gathers and creates the heat which causes it to rise. Much of that heat is transferred to the sexual glands, causing an increase in stimulation there. If one allows that energy to be expended frivolously in sexual indulgence, one loses a great portion of one's *Shakti*. But if it is conserved, it rises, and is absorbed into the body, resulting

in greater bodily vigor and luster, as well as greater mental power. This is a practice recommended for *brahmacharis* or *sannyasins* (monks). Married men and women, of course, are exempt from this kind of *tapasya*; for such as these, normal moderation is best.

(7)   *Longing for liberation:* Most important, for conserving and increasing the *Shakti*, is a strong aspiration toward, and longing for, liberation. Such aspiration is synonymous with the love of God, for such love is nothing more than a drawing of the heart toward the clarity and joy of absolute Truth. Such aspiration or love will draw the Grace of God, and will focus the energy upward toward the seat of Consciousness, and will be a strong counteractive to mental inertia and dullness.

According to the philosophy of Kashmir Shaivism, there are three different levels of spiritual practice; these levels, or methods (*upayas*), are: *anava upaya*, which is practice on the physical and sensual level; *shakta upaya*, which takes place on the mental, intellectual, level; and *shambhava upaya*, which engages the will and the intuition, and is on the astral or soul level. There is a fourth *upaya*, which is not really a practice at all, but an established awareness of the Self, and is therefore known as *anupaya*, or "no practice." This conceptual division can be simplified somewhat if we simply say that we exist on four levels: "the physical," "the mental," "the astral or soul-level" and "the spiritual." Our activities in pursuit of the Self take place on each of these progressively subtle levels, and become increasingly effective as we reach to increasingly subtler levels of activity.

Without doubt, we are all complexly constituted of body, mind, soul, and Spirit. Indeed, all is Spirit, but that Spirit manifests in a progressively more tangible manner as soul, as mind, and as body. According to the subtlety of our awareness, we identify ourselves most predominantly with one or another of these levels of our reality. Normally, we are aware of ourselves as a mixture of several of these elements; but one or another aspect of ourselves is usually a predominant focus. For example, the athlete focuses predominantly on his or her physical fitness, and measures his or her competency according to the abilities and qualities of the physical body. It would be foolish to say that the mind plays no part in such a person's awareness, but it is clear that much of the attention of that person's awareness is on their physical well-being and skills. This is true also of those people who labor in the so-called "lower" echelons of trades requiring physical exertion and manual dexterity. We see this body-orientation much more exaggerated, of course, in the animal realm, where physical instinct predominates to a much greater degree, and the mental realm is little developed.

The person who identifies predominantly with the mind gives less attention to the physical body, and more attention to the comprehension and structuring of ideas. Their focus is on exploring their understanding of ideas, mental task-accomplishments, and the comprehension of their world. They may be "intellectuals," or merely normal goal-oriented and career-oriented people. The more mentally developed may become writers, scientists, scholars, or technological experts; others comprise the vast majority of businessmen, teachers, white-collar workers, etc. Again, let me stress that, for most of us, there is a complex mixture of physical, mental, and soul-qualities at work in our lives, and none of these is omitted in our overall awareness; and yet, it is also certain that there is clearly a *predominant* focus on one or another of these aspects in each of our lives by which we may be "typed" in various ways.

The person who identifies predominantly with the soul is a person who has become opened to the subtler astral level of reality. Such persons are governed by a sense of the underlying unity of life, and strive to give expression to qualities of love, kindness, and compassion in their lives, with a strong sense of their purpose as a nurturing and inspiring presence in the world. Such people may become religious leaders, doctors, or crusaders for the social welfare. They are aware, not only of the tools they possess in the way of physical and mental abilities, but are motivated to use these God-given tools to benefit others and to lead the world toward peace and brotherhood. The individualized "soul" is that conglomerate of deeply ingrained qualities, evolved over many lifetimes, which makes up the character and purpose of an individual; and the person who identifies with the soul is one whose greatest emphasis is on perfecting the qualities of wisdom and love and on manifesting their own unique destiny in a way which will better themselves and all mankind. There is in the soul a clearer awareness of one's source in Spirit, and so with those who identify with the soul there is a strong desire to manifest that unifying Spirit, and to draw ever nearer to awareness of their own ultimate Being.

That ultimate Existent is the Spirit. The Spirit is that unmanifest Source from which all beings manifest. It is the unqualified Ground of all existence, which, in Itself, transcends all manifestation. It has been spoken of as pure Consciousness and Bliss; It has been spoken of as Brahman, God, or the supreme Self of all. It is that eternal Self with which the saints—the most evolved human beings—identify. They see that the body, mind, and soul are transient elements of their being, and that the One Spirit is their unchanging and eternal Identity. And they hold to their identification with that, paying but passing attention to the demands of body, mind, and soul. They realize that these have but a transient existence and will go on, by the operation of natural laws, but that they do not constitute their true Essence nor their purest happiness.

Those who identify with the Spirit, the eternal Self, find little to attract them to physical, mental or soul activity and accomplishments. Rather, they seek, and find, their greatest happiness and contentment in the awareness of their pure Being, beyond body, mind or soul. Such as these have no established place in the world; they are beyond the world of other men and women. Their vocation is to live in close union with God, and, though they may be regarded as monks, renunciants, or simply as societal outcastes, they serve as emissaries of the Divine. They act, to be sure. They are not without thoughts. Their souls have become expanded to include all souls in the One in whom they subsist, and their actions and thoughts derive from their Identity as the all-inclusive One; and, though their value is not recognized by the people of the world who are busily engaged in their own self-involved thoughts and activities, such people give clarity and light to the world, and serve as magnets to draw others to the all-gratifying Truth which exists within them all.

In the ancient world of Vedic India, this rudimentary division of peoples was translated into a set of classes or "castes," and was recognized as a natural fact of life; but as time went on, these stratifications of society became calcified into rigid air-tight compartments into which one was born and from which there was no escape. What had been an observation of natural evolution became an inflexible societal stratification based on racial and familial association. This was, of course, a distortion and corruption of what had been a keen observation of the varied levels of human awareness. That observation—that people do indeed fall roughly into several broad "types" according to the evolution of their awareness—remains, nonetheless, a valid one.

Recall how, in the Indian epic, the *Ramayana,* Rama, an incarnation of Vishnu, asks Hanuman, his monkey-servant (representative of the individualized soul), "How do you regard me?" And Hanuman replies, "When I regard myself as the body, I am your servant; when I regard myself as the mind, I am a part of you; and when I regard myself as the spirit, the Self, you and I are one." Note that Hanuman's realization became more subtle and closer to the nondual Truth as he went from identification with the physical body to the mind, and from the mind to the Spirit. From the perspective of Kashmir Shaivism, all our efforts toward personal growth and Self-realization manifest on one or another of these levels of reality. At the grossest level, we identify with the body; we regard ourselves as the servant of God, as His instrument; we perform physical acts: acts of service, ritual worship, Hatha yoga postures, the sounding of mantras, etc. These are necessary and

The Divine Universe

beneficial practices, but they are at the gross physical level only; we must go deeper toward the subtle if we are to reach God.

The next level of activity is the mental. Here, we perform many practices: we study the scriptures and other writings of the realized saints; we do mental worship, such as prayer, or the mental repetition of the name of God; we continually attempt to refine our understanding, and remind ourselves inwardly of the truth of the Spirit. And here, at this stage where we identify with the mind, we come to regard ourselves as a spark or a ray from the one Sun, which manifests and illumines the world. All is seen as God, and we are a part of Him.

Then, on the soul level, the activity is very subtle; we may also call it the level of consciousness. It is simply the constant alertness to reject any obscuration of conscious awareness. It is the jealous guarding of the pure Consciousness that is the witness, the Self. At that level, there is no duality of I and Thou, mine and Thine; there is only *I AM*. Notice that each one of these levels of activity leads to the next, subtler, level. For example, when you do physical acts of service, or worship, this brings with it the mental level of service or worship, as our concentration deepens. Or, if we repeat the name of God on the physical level, such as when we chant aloud, that physical repetition brings with it, by sympathetic resonance, the mental awareness of the name, and we find that we're repeating the mantra on the mental level as well. The idea, of course, is for our worship, our prayer, our meditation, to reach to deeper and deeper levels of subtlety, becoming a transforming force to recreate us at the spiritual level.

Practice at the mental level is superior, of course, to mere physical action, because it is by the transformation of our mind that we truly become transformed into Divine beings. As Krishna said to Arjuna in the *Bhagavad Gita,* "The Self is realized by the purified mind!" This is also what Jesus taught when he explained that it was the pure in heart who would see God. Also, we have seen what great emphasis is placed on the mental practice of Self-knowledge by the great Shankaracharya, who said, "The practice of knowledge thoroughly purifies the ignorance-stained mind, and then that [intellectual] knowledge itself disappears, just as a grain of salt disappears in water."

Shankara's analogy can be easily understood by one whose concentration on the knowledge, "I am pure Consciousness," leads the mind, through concentrated effort toward understanding, and eventually to perfect mental quietude, and the direct experience of pure Consciousness. Through one-pointed concentration on this one thought, "I am not merely this body, this mind; I am the Absolute; I am pure Consciousness," one goes beyond thought and attains the thought-free state. It is in this way that the mental practice leads to the subtler level of spiritual practice. The story of king Janaka and Ashtavakra is a good illustration of this:

King Janaka was sitting one day on the riverbank, repeating his mantra aloud. In a loud, powerful voice, he repeated over and over *So-ham, So-ham, So-ham;* "I am That! I am That! I am That!" Then, along came his guru, Ashtavakra, who sat on the opposite bank. Observing that king Janaka was involved in the physical practice of mantra-repetition, with maybe a touch of mental practice thrown in, Ashtavakra decided to elevate king Janaka's practice. So he began to shout aloud, "This is my water bowl! This is my staff!" And, as he did so, he alternately lifted each of the items mentioned. Ashtavakra continued this for a long time, shouting at the top of his voice, "This is my water bowl! This is my staff!"

Soon the king's mantra-repetition was disturbed and he quickly became annoyed. Finally, he could take it no more, and he shouted across to Ashtavakra, "Hey, why all this racket? I know those things belong to you; who says they're *not* yours?" And Ashtavakra shot back, "And who says you are *not* the Self?" Immediately king Janaka's mind ceased its activity and became absorbed in the silent awareness that he *was* the Self, and didn't need to go on engaging his lips or his mind in repeatedly asserting it. In other words, by the grace of his guru, his mental practice merged into the soul's awareness of its identity with the Self.

This practice does not call into play either the body or the mind, but rather what we would call simply, "the will." It is the practice of keeping a willful check on the impulses of the mind, and a willful retention of pure awareness, with a sense of identification with the one all-pervasive Consciousness. It is, in other words, a direct soul-awareness through the effort of will. In its highest stage, this subtle practice becomes no practice at all. It simply remains spontaneously, habitually. It is the state of consciousness which the Zen Buddhists call the state of "No-mind," which Vedantists refer to as *sahaj samadhi,* "the natural state of unity," and Kashmir Shaivites refer to as *anupaya.*

To explain how one level of practice leads to a subtler level, let's take, as an example, the practice of mantra repetition. You may begin by just repeating it on the physical level. And, even on this level, the sound-vibrations have a certain effect on you, instilling peace and a sense of well-being. Then, you begin to reflect on its meaning. Now, it is no longer just a sound; it's a meaningful thought: *So-ham.* The mind translates the sound into "I am That; I am the one Self." That is the mental practice. You repeat the mantra on the mental level with an awareness of its meaning. Then, as you begin to sense the reality of it, as you begin to experience it, you transcend the mantra, and hold yourself poised in the thought-free state. That's the level of soul-awareness, and is very close to the awareness of Spirit, or the Self. When, eventually, this awareness deepens, one loses all sense of body, mind, or soul, and, transcending all practice, becomes immersed in the awareness of the Self.

Now, to make all this really clear, I'm going to give you some sample practices from each of these three levels. And, to do that, I'm going to use an ancient scripture from the tradition of Kashmir Shaivism, one of the *agamas*, called the *Vijnana Bhairava*. "Bhairava" is another name for Shiva, the Lord, the Self. And "Vijnana" is the conceptual knowledge or expression of the Self. This expression takes the form of a dialogue between Shiva and his consort, Shakti.

In this imaginary dialogue, Shakti asks Shiva to explain His true nature and the practices by which he can be known; and Shiva then details 112 different practices, utilizing those from each of the three levels we've discussed. First, we'll hear of some of the physical practices, some of which have to do with the subtle breath, the *prana*, or the visualizing of inner lights and sounds. Listen to some of the practices Shiva recommends to Shakti. You might like to try them out as I mention them to you:

> The breath is exhaled with the sound, *Ham*, and inhaled with the sound, *Sah*. Thus, the individual soul always recites the mantra, *Hamsah* (or *So-ham*, "I am That!"). [6]

> *Prana* goes upward (with the inhalation), and the *apana* goes downward (with the exhalation). This is the expression of the creative Shakti. By becoming aware of the two places where each originates, experience absolute fulfillment. [7]

> There is a momentary pause, when the outgoing breath has gone out, and there is a momentary pause when the ingoing breath has gone in. Fix your mind steadily on these places of pause, and experience Shiva. [8]

> Always fix your mind on those places where the breath pauses, and the mind will quickly cease its fluctuations, and you will acquire a wonderful state. [9]

In the *Bhagavad Gita (4:29)*, Krishna says, "Some yogis, devoted to *pranayama* (the control of the *prana*), offer as sacrifice the outgoing breath into the incoming breath, and the incoming into the outgoing, restraining the course of both." It is this very practice that is being spoken of here in the *Vijnana Bhairava*, which goes on to say:

> When the ingoing breath merges with the outgoing breath, they become perfectly balanced and cease to flow. Experience that state and realize equality. [10]
>
> Let the breath remain balanced, and let all thoughts cease; then experience the state of Shiva. [11]

That's enough practices on the physical level; let's move on to the mental practices. Here, we enter into the realm of ideas. These practices deal entirely with formulated intellectual knowledge. Shiva says to Shakti:

> Concentrate your mind on whatever gives you satisfaction. Then experience the true nature of supreme satisfaction. [12]
>
> Meditate on yourself as a vast, cloudless sky, and realize your true nature as Consciousness. [13]
>
> Becoming detached from the awareness of the body, meditate on the thought, "I am everywhere!" and thus experience joy. [14]
>
> Hold this thought in your mind: "All the waves of the various forms in this universe have arisen from me—just as waves arise from water, flames arise from fire, or rays from the Sun. [15]
>
> Contemplate with an unwavering mind that your own body and the whole universe are of the nature of Consciousness, and experience the great awakening. [16]
>
> Contemplate your body and the whole universe as permeated with Bliss. Then experience yourself as that Bliss. [17]

Okay. Now we come to the practices involving the soul; these are at a yet subtler level of consciousness. Here, you don't have to think at all. You need only to become aware, focusing on that clear, thought-free awareness that is your soul, an individualized manifestation of the Self. Shiva says:

> Observe the arising of a desire. Then immediately put an end to it by reabsorbing it into That from which it arose. [18]

What are you when a thought or desire does *not* arise?
Truly, the one Reality! Become absorbed in and identified with
That. [19]

When a thought or desire arises, detach yourself from the object
of thought or desire, and witness the thought or desire as a
manifestation of your Self, and thus realize the Truth. [20]

The same conscious Self is manifest in all forms; there is no
differentiation in It. Realize everything as the same One, and rise
triumphantly above the appearance of multiplicity. [21]

When under a strong impulse of desire, or anger, greed, infatuation,
pride, or envy, make your mind steady and become aware of the
Reality underlying the mental state. [22]
Perceive the entire universe as a magic-show, or as forms painted
on a canvas, or as so many leaves on a single tree; and becoming
absorbed in this, experience great happiness. [23]

Leaving aside your own body for the time being, contemplate your
Self as the consciousness pervading other bodies, and thus become
all-pervasive. [24]

Free the mind of all supports, without and within, and let no
thought-vibration take form. Then the self becomes the supreme
Self, Shiva. [25]

At the onset or culmination of a sneeze, or at the moment of fright,
or deep sorrow, or at the moment of a sigh, or while running for
your life, or during intense fascination, or extreme hunger, become
aware of Brahman. [26]

What cannot be objectively known, what cannot be held in the
mind, that which is empty, and exists even in non-existence:
contemplate That as your Self, and thus attain realization of Shiva.[27]

Meditate on yourself as eternal, all-pervasive, the independent Lord
of all; and thus attain That. [28]

The Perennial Philosophy

About *anupaya*, the ultimate state beyond all practice, there is really nothing one can say. It is the reversion of the soul to its universal Source. In such a state, one is on a pathless path, beyond the bodily, mental or astral levels. Immersed in God-awareness, there is no more striving, for there is nothing more to attain. There is no action, no thought, no individual awareness. There is only the pure Bliss of the Self.

Here is what Jnaneshvar, the 13th century yogi, says in his *Amritanubhav* about such a state:

> One who has attained this wisdom may say whatever he likes; the silence of his contemplation remains undisturbed. His state of actionlessness remains unaffected, even though he performs countless actions. Whether he walks in the streets or remains sitting quietly, he is always in his own home. His rule of conduct is his own sweet will. His meditation is whatever he happens to be doing. [29]

Such a knower of the Self lives in perfect freedom. You too, by utilizing all these practices—of the body, the mind and the soul—can attain eventually to that (fourth) state. As you meditate, just sit quietly; let the mind be still and become aware of the Self. If you can't do that immediately, then take the help of the mantra, the name of God, the name of the Self. Reflect on its meaning. Identify with that One. And if you cannot do that, at least practice on the physical level: repeat the mantra with the in-breath, and again with the out-breath. Let it carry you to the awareness that you and your beloved God are one.

Notes:

1. Shankara, *Vivekachudamani*, III:16; Swami Prabhavananda and C. Isherwood, *Shankara's Crest-Jewel of Discrimination*, Hollywood, Vedanta Press, 1947; pp. 70-71.

2. Jnaneshvar, *Amritanubhav*, 7:165, 166; Swami Abhayananda, *Jnaneshvar: The Life And Works, etc.*, Olympia, Wash., Atma Books, 1989; p. 186.

3. *Ibid.*, 7:170, 171; p. 187.

4. *Ibid.*, 7:235-237, 240; pp. 193-194.

5. Jesus, *The New Testament of the Bible:*, *Luke*, 18:18; *Matthew*, 19:16.

6. *Vijnana Bhairava,* 155.

7. *Ibid.,* 24.

8. *Ibid.,* 25

9. *Ibid.,* 51.

10. *Ibid.,* 64.

11. *Ibid.,* 26.

12. *Ibid.,* 74.

13. *Ibid.,* 92.

14. *Ibid.,* 104.

15. *Ibid.,* 110.

16. *Ibid.,* 63.

17. *Ibid.,* 65.

18. *Ibid.,* 96.

19. *Ibid.,* 97.

20. *Ibid.,* 98.

21. *Ibid.,* 100.

22. *Ibid.,* 101.

23. *Ibid.,* 102.

24. *Ibid.,* 107.

25. *Ibid.*, 108.

26. *Ibid.*, 118.

27. *Ibid.*, 127.

28. *Ibid.*, 132.

29. Jnaneshvar, *Amritanubhav,* 9:20, 21, 31, 34; S. Abhayananda, *Op. Cit.;* pp. 207, 208.

18.

OX-HERDING

There is a set of ten picture-drawings in the Chinese (*Ch'an*) Buddhist tradition, called "The Ox-Herding Pictures," which tells the story of the spiritual journey in a parable form. There were originally eight pictures, created by some nameless artist of the Taoist tradition many centuries prior to the establishment of Buddhism in China around the 6th century C.E; but the eight pictures were extended to ten by a 12th century Chinese Buddhist named Kakuan, who also wrote verses to go with each picture. We can easily understand the meaning of these cryptic pictures and verses from our perspective based on the Vedantic and Kashmir Shaivite teachings cited in the previous Essays. Bear in mind that there are not many teachings represented by the many different religious traditions, but only one. They are all the same teaching. It is true that the teachings of the knowledge of the ultimate Unity experienced directly through interior realization had its early expression in India; the Upanishads form one of the earliest known expressions of that knowledge. But, of course, the experience is universal; men and women everywhere have experienced the unitive Self, and spoke of it in their own language and in their own way. And this knowledge spread, along with the various ways of talking about it.

Buddhism had its growth from Hinduism—just as Christianity was a descendent of Judaism, insofar as its concepts and terminology were derived from a pre-existent Judaic culture. The Buddha, who grew up in a Hindu culture, long prior to the full development of Advaita Vedanta philosophy, simply phrased his knowledge of Unity in a new and unique manner; but it was not a new Truth he taught. Today, Buddhism is a concensus of many teachings, none of which we may be certain originated with the Buddha since he wrote nothing, and the written texts purporting to be his teachings were gathered together centuries after the Buddha was gone. Then, as the Buddhist teachings spread to Tibet, China, and Japan, they took on the character of the cultural and linguistic traditions of those countries. Chinese Buddhism is therefore unlike its Indian counterpart in its style and manner, but not dissimilar in its essence. The Truth to which Ch'an and Zen points is the same Truth to which the Buddha pointed, the same Truth to which the Upanishads pointed. The Reality experienced is the same for all, but there is room for immense diversity in the expression of it. Each path, though unique, leads to the same, single, destination. This will become clearer as

The Perennial Philosophy

we talk about these ten Ox-Herding pictures of the Ch'an Buddhists. The language and style of these ancient Chinese texts is somewhat different from that of Vedanta, but we will see that the message of Unity is the same for all.

(1) The first picture, "The Search For The Oxen," shows a young man searching through the woods for the oxen. He's standing by a riverbank, wondering which way he should go. The verse of Kakuan, which accompanies this picture, reads:

> In the pasture of this world, I endlessly push aside the tall grasses in search of the oxen. Following unnamed rivers, lost upon the interpenetrating paths of distant mountains, my strength failing and my vitality exhausted, I cannot find the oxen. I hear nothing but the locusts chirring through the forest at night.

Comment: *First of all, what is this ox the young man is searching for? The oxen represents ultimate knowledge. It is this we are all seeking. Whether we call it "God," "Brahman," the "Tao," the "universal Mind," or simply the "Truth," we possess an inherent longing for it in our hearts. And, beyond the obscuring "tall grasses" of this world, we are seeking to catch a glimpse of the Truth of existence. This is the elusive oxen of our young man's search. And in his long search, he has followed endless philosophies, labyrinthine twists of speculation and logic, and he has only become more confused, more desperate, more weary of the search; and he hasn't a clue as to which way to turn. He has heard no guiding voice of God in the wilderness; he hears only the mocking sounds of the chattering crickets and locusts in the darkness.*

(2) The second picture, called "Discovering The Footprints," shows the young man running alongside the hoof-prints of the oxen, which lead off into the distance. He carries a rope and a whip with which to capture his prey. The verse accompanying this picture reads:

> Along the riverbank under the trees, I discover footprints! Even under the fragrant grass I see his prints. Deep in remote mountains they are found. These traces no more can be hidden than can one's nose, when looking heavenward.

Comment: *Only in the remote mountains, far from the activities of common men and common pursuits, only on the high peaks of thought and yearning, does one find traces of the Eternal. Then, His signs are evident even in a blade of grass. When the eyes are cleared, His traces are evident everywhere we look, planer than the nose on our face! But note that, in the parable, the young man is objectifying the truth (the oxen), imagining it to be something "other" than himself. He is still immersed in duality, still seeing from the standpoint of a limited "I" who is seeking a "Thou" out there.*

(3) The third picture, entitled, "Perceiving The Oxen," shows the young man having found the oxen, getting ready to capture it with his rope made into a halter. The verse accompanying it states:

> I hear the song of the nightingale. The Sun is warm, the wind is mild, willows are green along the shore. Here, no ox can hide! But what artist can draw that massive head, those majestic horns?

Comment: *Now that he has comprehended the Divine, Its beauty and glory is evident in the pleasant sensations of the world. Our young man feels the closeness of the Divine, and feels he is on the verge of capturing his prize. It is almost within his grasp. The truth is now self-evident! How could It hide from the man of clear vision! And yet, It is beyond description or conception; It is so great, so vast, who could do It justice by description or art? But note: so long as the Truth which the young man conceives remains "other" than himself, so long is his Truth a mere mirage, a product of his own thought.*

(4) In the fourth picture, entitled "Catching The Oxen," we see the young man, having gotten his rope around the hind leg of the ox, holding on for dear life, as the oxen struggles to get away. The verse says:

> I seize him with a terrific struggle. His great will and power are inexhaustible. He charges to the high plateau far above the cloud-mists, or he goes to stand in an impenetrable ravine.

Comment: *Now it has become clear that this "Truth" which our young man has got hold of is not ultimate Truth at all, but his own mind's creation. It is only thought that he has roped; he is chasing only his own mind! It takes him up to the heavens on the subtlest perception, only to careen into the lowest, animal, depths. It drags him up and down, around and around. Who has ensnared who? This "terrific struggle" is the* sadhana *of taming one's own mind. But the mind has an inexhaustibly powerful will. It tosses him about like the wind, while he hangs on for dear life.*

(5) The fifth picture, called "Taming The Oxen," shows the young man leading the oxen along the road peacefully by the rope in his hand, with his whip in the other. The verse states:

> The whip and rope are necessary; else he (the ox) might stray off down some dusty road. Being well-trained, he becomes naturally gentle. Then, even when unfettered, he obeys his master.

Comment: *The whip is the prodding of the will; the rope is the* mantra, *or* name, *by which one keeps a firm grip on the wayward mind. The mind untamed will surely dart off down every impure path; but, once kept under tight control, it learns to be quiet and pure, peaceful and gentle, attentive to its master, even when the controls are relaxed. This reminds us of the teaching of the* Bhagavad Gita, *wherein Arjuna complains to Krishna:* "O Krishna, the mind is inconstant; in its restlessness, I cannot find any rest. This mind is restless, impetuous, self-willed, hard to train; to master the mind seems to me as difficult as to master the mighty winds."

To which, Krishna answers: "The mind is indeed restless, Arjuna; it is indeed hard to train. But by constant practice and by freedom from passions the mind can, in truth, be trained. When the mind is not in harmony, Divine communion is hard to attain; but the man whose mind is in harmony attains it, if he has knowledge and persistence."[1]

(6) The sixth picture, "Riding The Oxen Home," shows the young man astride the oxen, riding along peacefully, playing his flute. And the verse, which accompanies it, states:

> Mounting the ox, slowly I return homeward. The voice of my flute intones through the evening. Measuring with hand-beats the pulsating harmony, I direct the endless rhythm. Whoever hears this melody will join me.

Comment: *This is a very interesting verse. The young man is doing his* sadhana, *his practices, meditating with a quiet mind (oxen), and progressing toward his true home in the Self. The sound of his flute is heard through the night. There are frequent allusions in the literature of yoga to the stage of* sadhana *in which one begins to hear inner sounds emanating from within. The sound of the flute is one of the most common referred to. It is mentioned not only in the yogic tradition of India, but also in the ancient Taoist tradition of China. Yogis speak of this inner music as* nada. *In the Sikh tradition, it is spoken of as* naam, *or* shabd. *In the yogic texts, this unstruck sound (*anahat nada*) is spoken of as a spontaneous occurrence, which fills the inner ear with music of different kinds, seemingly produced by different instruments, such as the flute, the vina, the mrdung. It is apparently to this that our poet refers. He says, "Whoever hears this melody will join me," meaning, they will be on a par with him, and like him, will reach the Destination. For it is said that this* nada *is like the rosy color before the dawn; one who listens to it with concentration will be led to the ultimate experience of Unity.*

(7) The seventh picture, "The Transcendence Of The Oxen," is, of course, the transcendence of the mind. It shows the young man sitting all alone by his hut peacefully. The oxen is nowhere in sight. The verse states:

> Astride the oxen, I reach home. I am serene. The oxen too can rest. The dawn has come. In blissful repose, within my thatched dwelling I have abandoned the whip and the rope.

Comment: *"Astride the oxen, I reach home": mastering his mind, he has attained his final destination. He has conquered the unconquerable mind; "the oxen too can rest now." No more need of whip and rope. No more discipline, such as mantra-japa, or prayer, or worship. The mind is silenced; it does not show itself. All ignorance has disappeared. Note too that it is not the oxen—the original "object" of the quest—who remains alone; but it is the "subject" only who is left. The object has proven illusory.*

(8) The eighth picture, the final one in the Taoist version, is called "Both Oxen and Self Transcended." In this picture, there is nothing at all. It is blank. Nothing. The verse beneath states:

> Whip, rope, person, and oxen—all are merged in the Featureless. This heaven is so vast no message can stain it. How may a snowflake exist in a raging fire? Here are the footprints of the patriarchs.

Comment: *Here is portrayed the fact that, without an object, the subject also disappears; and what is left is the Formless. The process of bifurcation has been transcended in* nirvana, *in* samadhi. *Just as a salt-doll cannot exist when immersed in the ocean, or a snowflake cannot exist in a raging fire, neither can an imaginary individuality exist in the ocean of Oneness. Neither is there any trace (footprints) of those who have passed before. Here, there is no other reality but the One, the Featureless Ground. This is the so-called "union of the soul and God," "the mystical marriage," "the experience of Unity."*

(9) The ninth picture, which was added by Kakuan along with the tenth, shows a nature scene: a willow tree bending over a babbling brook wherein fishes play and above which the hummingbirds hover. Leaves fall from the tree into the stream, indicating the changing of the seasons. The verse reads:

> Too many steps have been taken returning to the root and source. Better to have been blind and deaf from the beginning! Dwelling in one's true abode, unconcerned with what is without—the river flows tranquilly on and the flowers are red.

Comment: *The Taoist who originally framed this little story stopped at the eighth picture. All was reduced to the transcendent One; the division of subject and object was no more. But the Buddhist, Kakuan, found more to say and added two more pictures. From the viewpoint of Ch'an or Zen Buddhism, it was essential to do so. These down-to-earth Buddhists do not care for the apparent world negating of the Indian and Taoist metaphysics; nor even for the "Emptiness" of their own Buddhist tradition as it was taught elsewhere. The seers of both the Chinese and Japanese Buddhist tradition seem extremely interested in bringing the realization of Unity to its practical conclusion in the world, which must continue to exist as experience. Here in this verse, the poet says, "It would have been better if I had been born blind and deaf, instead of having to go through all the world-negating I went through to reach the awareness of the Absolute. But, now that I have realized the eternal Truth, I know that, while I transcend the world of form, still I am here watching the river flow serenely by, and I am seeing the redness of the roses. This life of samsara is not other than the one Consciousness. Samsara and nirvana are the same."*

The poet is alert to the fact that he is both free from and contained in the world; that he is both the transcendent One and the immanent manifold appearance; that (in the words of Vedanta) he is both Brahman and Maya, both Shiva and Shakti. After all, if a small cell within the body of a man realized that it had no separate identity, but was in reality the whole man who contained in himself billions of cells; still, he would have to continue to live and function as a separate cell. Only by living and acting within his larger self in a way appropriate to the needs of a cell would he be able to benefit his larger self. In the case of the Self-realized sage, the knowledge of his universal Identity makes him free; yet he must continue to live and act in a meaningful way as a limited entity. He lives in the world, and is at the same time above it, as the transcendent Self.

The Divine Universe

(10) The tenth picture, "In The World," shows the young man, now grown older, mingling in the marketplace with the fruit sellers, as a wizened old sage, staff in hand, looking like a fat, jolly, Buddha. The verse with it states:

> Barefooted and bare-breasted, I mingle with the people of the world. My clothes are ragged and dust-laden, yet I am ever blissful. I possess no magic to extend life; yet, before me, the dead trees put forth blossoms.

Comment: *The young man is now older and wiser. He seems to others a poor fool, yet he is eternally blissful. The world around him appears as a shimmering lake of jewels. All is alive with Consciousness, and the energy of his Consciousness enlivens all about him. He is free of all worldly endeavor, yet he wanders about joyfully, finding happiness everywhere. This is the familiar picture of the* Avadhut, *the blissful sage of the Vedantic tradition, who has completed his* sadhana *and has nothing further to accomplish. We meet with him in the* Avadhut Gita, *where he is depicted in this way:*

> A patched rag from the roadside serves as a wrap
> To the Avadhut, who has no sense of pride or shame.
> Naked, he sits in an empty shack,
> Immersed in the pure, stainless, bliss of the Self.
> Free from bondage to the fetters of hope,
> Free from everything, he has thus attained peace.
> He is the stainless One, the pure Absolute.
> For him, where is the question of being embodied or bodiless?
> Where is the question of attachment or non-attachment?
> Pure and unpartitioned as the infinite sky,
> He is, himself, the Reality in Its natural state.
> As a *yogi* (seeker of union) he is beyond union and separation;
> As a *bhogi* (worldly enjoyer), he is beyond enjoyment and non-enjoyment.
> Thus, he wanders leisurely, leisurely,
> While in his mind arises the natural bliss of the Self. [2]

Listen also to the way the Maharashtran saint, Jnaneshvar, spoke in the 13th century about the state of one who has reached this final liberation:

> One who has attained this wisdom may say whatever he likes; the silence of his contemplation remains undisturbed. His state of actionlessness remains unaffected, even though he performs countless actions.
>
> ... Even one who has attained wisdom may appear to enjoy the sense-objects before him, but we do not really know what his enjoyment is like. If the moon gathers moonlight, what is gathered by who? It is only a fruitless and meaningless dream!
>
> ... Sweeter even than the bliss of liberation is the enjoyment of sense-objects to one who has attained wisdom. In the house of devotion that lover and his God experience their sweet union. Whether he walks in the streets or remains sitting quietly, he is always in his own home. He may perform actions, but he has no goal to attain. Do not imagine that if he did nothing, he would miss his goal. ... His rule of conduct is his own sweet will. His meditation is whatever he happens to be doing. The glory of liberation serves as a seat-cushion to one in such a state.
>
> No matter where he goes, that sage is making pilgrimage to God. And if he attains to God, that attainment is non-attainment. How amazing! That in such a state, moving about on foot and remaining seated in one place are the same. No matter what his eyes fall upon at any time, he always enjoys the vision of God. [3]

All the great Enlightenment traditions speak of this synthesis of the Eternal and the temporal, the Divine and the mundane, as the final liberation. Such a sage is known as a *jivanmukta*—one who is free while living. He lives in the world (for where else would he live?); but the knowledge of his eternal Self has freed him from identification with the apparent limitations of embodiment. He is free. We too can attain such a state. Most of us are still at one stage or another along the way to perfect freedom, perfect awareness. We keep on learning and practicing and doing what we must to tame our unruly minds. In this way, we lift our consciousness to greater and greater heights, till finally we know our boundless, joyful and eternal freedom. May

The Divine Universe

we all, this very day, begin to taste a little of the bliss of our true, carefree and omnipresent Self.

Notes:
1. Krishna, *Bhagavad Gita,* 6:33-36.

2. Dattatreya, *Avadhut Gita,* VII:1, 3, 4, 9; Swami Abhayananda, *Dattatreya: Song of The Avadhut,* Olympia, Atma Books, 1992.

3. Jnaneshvar, *Amritanubhav,* 9:20, 21, 25, 30-32, 34, 53-55; Swami Abhayananda, *Jnaneshvar: The Life And Works, etc.,* Naples, Florida, Atma Books, 1989; pp. 207-209.

19.

THE APPEARANCE OF DUALITY

It is well known that the Self of man and the ultimate and transcendent Reality known as God are not two. This is the perennially acceptable view of "Nonduality". But it must also be acknowledged that there is an *apparent* duality which has a certain phenomenal reality to it as well. For, during the "mystical experience" one experiences a noumenal and eternal 'I' who manifests this universe in which lives a phenomenal and temporal 'I'. The 'I' is the same, yet different. The difference between the two 'I's is that the eternal one projected Himself as the temporal one into this world of time and space; the temporal one did not project himself into eternity.

So, God, by His very *projection* of this temporal universe, establishes an *apparent* duality for those living within this projection. This is not difficult to understand: If there is a dreamer and his dream, there *appears* to be two. But are there really two? The truth is that there is still only one; the other is only an imagination, and though the consciousness in the dream seems to be an 'other', it is in fact the consciousness of the dreamer. But some would argue: still, the other *exists* as a phenomenon, and therefore constitutes a second. It is a question of perspective, is it not? At least we may be certain that, once the dreamer awakes and the dream is no more, then only one remains. The Nondualist would no doubt remark that there was *always* only one.

We dream-images enclosed within this illusory universe of time and space, are similarly "phenomena", and therefore *appear* to exist. And so, as *images* of God (who is our *true* Self), we regard God as separate, 'other'. For, while *we* are enclosed within the world of time and space which is His projection made of His Consciousness, *He* is nonetheless entirely beyond it. He is the eternal Mind that projects this space/time continuum, this form-filled world, as a construct of thought. He is indeed the Consciousness which animates us and which lends us consciousness. He is our very Self; He is the one and *only* Reality. But it is not wrong to acknowledge the *apparent* Duality which He brings to pass in the act of projecting this world of beings within Himself.

Though, ultimately, when we pass from space-time to the unlimited Reality, we shall recognize the eternally inseparable oneness of God and our Self; nonetheless, while living as separate beings within this worldly illusion, it is quite understandable if we call out to Him as though He were separate,

The Divine Universe

or 'other', just as dream figures might call out within themselves in an effort to contact the dreamer, who is indeed their own essence, a one who becomes an *apparent* two.

Some hold exclusively to the eternal truth of unity, declaring their single and only identity to be 'the One'; these are the *jnanis* (or "knowers"). Others, acknowledging the *apparent* duality between themselves and God, worship the One as other than themselves, as the Exemplar of which they are mere images. These are the *bhaktas* (or "lovers") And both are perfectly correct and valid pathways to the realization of God, the knowledge of the eternal Self. The *jnani* says, "I am That"; the *bhakta* says, "O Lord, Thou alone art!". And both arrive at the selfsame realization of the Real.

'And what of the apparent duality of body and spirit?' we may wonder. We all know what Descartes thought about it. But I would ask, 'Have you ever seen ice cubes floating in water? Are they two things or one?' There seems to be two different substances, since each is clearly separate from the other; but no, it is one substance in two different states. When I was immersed in the unitive vision, I wondered "Where is the temple (of the body)? Which the imperishable, which the abode?" For there was to be seen no separate body-temple with an imperishable soul within! There was no division to be found at all. All is Consciousness-Energy in this dream! And all of it is imperishable. It is only the various shapes that are so changeable, so very perishable; but the Essence is one.

Think of your own dream-creations! Is your dream-character divided into a consciousness and a body-form? No. It is one thing: the form and its limited self-consciousness are one projected creative mind-stuff. Likewise for us here on earth. We live and move and have our being within the Mind-stuff of God. It is His drama, and He is the Self-consciousness of each of us. When ultimately we awake, we shall know the Source of all selves, the Source of all forms; we shall know that we were, are, and ever shall be, the One who lives in eternal bliss.

But what of the separation between the 'soul' and the body at death? It seems quite certain that consciousness withdraws from the body when the heart stops beating, that consciousness and energy then go their separate ways. And that seems to imply a real, absolute, duality. But it is just the magic of the One. Think of what happens when you wake from a dream: Your own consciousness of Self remains even when the dream vanishes. Who you thought you were in the dream is seen to have been an illusory identity; but *You* remain. The dream scenery is vanished too. Where did it go? It never really was. It too was only your own consciousness, *appearing* as form. Likewise, in this universe, matter is consciousness appearing as energy, appearing as form.

The Perennial Philosophy

The universe itself is occurring *as a whole* within the one Consciousness. It is an integral dream-like phenomena. He is always One, even while projecting the universal dream with His Consciousness-Energy. When each of the dream-like images awakes, they awake to the One. Then, at the end of the universal 'dream', all forms revert to Energy, which ceases its transformations and merges into the one Consciousness. Consciousness ceases its play, resolving quietly into Itself. They were never two; they are merely twin aspects of His projective Power. The Supreme Consciousness will rest now, prior to projecting once again an apparent universe of conscious forms, another seeming duality upon His oneness.

Keeping this unity-in-duality, or duality-in-unity, in mind, please reconsider the remarkable text from the Gnostic seer, Simon Magus (*fl. ca.* 40 C.E.), entitled *The Great Exposition*, which so ably explains the apparent duality within the Nondual reality:

THE GREAT EXPOSITION [1]

There are two aspects of the One. The first of these is the Higher, the Divine Mind of the universe, which governs all things, and is masculine. The other is the lower, the Thought (*epinoia*) which produces all things, and is feminine. [2] As a pair united, they comprise all that exists.

The Divine Mind is the Father who sustains all things, and nourishes all that begins and ends. He is the One who eternally stands, without beginning or end. He exists entirely alone; for, while the Thought arising from Unity, and coming forth from the Divine Mind, creates [the appearance of] duality, the Father remains a Unity. The Thought is in Himself, and so He is alone. Made manifest to Himself from Himself, He appears to be two. He becomes "Father" by virtue of being called so by His own Thought.

Since He, Himself, brought forward Himself, by means of Himself, manifesting to Himself His own Thought, it is not correct to attribute creation to the Thought alone. For She (the Thought) conceals the Father within Herself; the Divine Mind and the Thought are intertwined. Thus, though [they appear] to be a pair, one opposite the other, the Divine Mind is in no way different from the Thought, inasmuch as they are one.

Though there appears to be a Higher, the Mind, and a lower, the Thought, truly, It is a Unity, just as what is manifested from these two [i.e., the universe] is a unity, while appearing to be a duality. The Divine Mind and the Thought are discernible, one from the other, but they are one, though they appear to be two.

[Thus,] ... there is one Divine Reality, [apparently] divided as Higher and lower; generating Itself, nourishing Itself, seeking Itself, finding Itself, being mother of Itself, father of Itself, sister of Itself, spouse of Itself, daughter of Itself, son of Itself. It is both Mother and Father, a Unity, being the Root of the entire circle of existence.

Notes:

1. Simon Magus, *Apophasis Megale* ("The Great Exposition"), quoted by Hippolytus of Rome, in *Refutatio Omnium Heresium* ("The Refutation of All Heresies"), VI.8; adapted from Roberts, Rev. A. & Donaldson, J. (eds), *The Ante-Nicene Christian Library*, Vol. VI; Edinburgh, T. & T. Clark, 1892; pp. 208-210. As this text is one of my favorites, it has been cited by me previously in Abhayananda, Swami, *History of Mysticism*, Olympia, Wash., Atma Books, 1987, 2000; p. 132; and again, with commentary, in Abhayananda, Swami, *Mysticism And Science*, Winchester, U.K., O Books, 2007; pp. 66-72.
2. For an extended discussion of the 'male-female' symbology, see the two book references cited above.

ღ   ღ   ღ

20.

NONDUALISM IN THE TEACHINGS OF JESUS [1]

Part One

Nondualism is a term applied to both religion and philosophy; and yet it must also be said that it refers to a state of awareness beyond both philosophy and religion. Truly speaking, *Advaita* is a description of and commentary on the nature of Reality as directly experienced in "the mystical vision." Only those who have actually experienced the Truth directly are able to speak authoritatively about It. And, the fact is, there have been many wise and pure-hearted men and women of every nationality and every philosophical and religious affiliation who have experienced the Truth. There are Christians who have experienced It, and Jews, and Muslims, and Hindus, and Buddhists, Neoplatonists, and so on. And so, we must include as part of the Nondualist heritage the teachings and writings of all those of various traditions who have directly realized the Truth and spoken of It.

Let us consider, for example, some of those Christians who taught the philosophy of Nondualism under the guise of Christianity. They are the seers, the mystics of the Church, who taught the path to God-realization, and who proclaimed the identity of the soul and God, and the indivisibility of the one absolute Reality. First among these, of course, is Jesus of Nazareth, called "the anointed one," or *Christos*, in the language of the Greeks. It is of his own mystical experience that Jesus spoke, a mystical experience which transcends all doctrines and all traditions, and which is identical for Christians, Muslims, Jews, and Vedantists alike. It is an experience of absolute Unity—a Unity in which the soul merges into its Divine Source, and knows, "I and the Father are one."

This knowledge is unacceptable in all orthodox religious traditions, however; and so, those, like Jesus, al Hallaj, Meister Eckhart, and many others who have experienced the Truth, are inevitably rejected by the religious traditions to which they belong. The religious tradition, which arose around the teachings of Jesus, in its turn, ironically, rejects its mystics as well. Nonetheless, down through the centuries, a few of the followers of Jesus also

The Divine Universe

experienced this Unity, by the grace of God, and spoke of It for posterity. Here, for example, is what the famous Christian mystic of the 13th century, Meister Eckhart, had to say about his own experience:

> As the soul becomes more pure and bare and poor, and possesses less of created things, and is emptied of all things that are not God, it receives God more purely, and is more completely in Him; and it truly becomes one with God, and it looks into God and God into it, face to face as it were; two images transformed into one. ... Some simple people think that they will see God as if He were standing there and they here. It is not so. God and I, we are one. [2]
> ... I am converted into Him in such a way that He makes me one Being with Himself—not (simply) a similar being. By the living God, it is true that there is no distinction! [3]

Or this, by the 15th century Christian Bishop, Nicholas of Cusa:

> Thou dost ravish me above myself that I may foresee the glorious place whereunto Thou callest me. Thou grantest me to behold the treasure of riches, of life, of joy, of beauty. Thou keepest nothing secret. [4]
> I behold Thee, O Lord my God, in a kind of mental trance, [5] ... and when I behold Thee, nothing is seen other than Thyself; for Thou art Thyself the object of Thyself, for Thou seest, and art That which is seen, and art the sight as well. [6]
> Hence, in Thee, who are love, the lover is not one thing and the beloved another, and the bond between them a third, but they are one and the same: Thou, Thyself, my God. For there is nothing in Thee that is not Thy very essence. [7] Nothing exists outside Thee, and all things in Thee are not other than Thee. [8]

Or listen to this, by the 16th century Christian monk, St. John of the Cross:

> What God communicates to the soul in this intimate union is totally beyond words. In this transformation, the two become one. [9]
> ...The soul thereby becomes Divine, becomes God, through participation, insofar as is possible in this life.
> ... The union wrought between the two natures, and the communication of the Divine to the human in this state is such that even though neither changes their being, both appear to be

God.[10] ... Having been made one with God, the soul is somehow God through participation.

This is the Truth revealed in "the mystical vision," the Truth that mystics speaks of as "Nonduality." While some Christians interpret Saint John's words to indicate that "the mystical experience" of Unity is an aberration, a gracious revelation to the soul of the nature of God, rather than a revelation of the underlying unity of the soul and God, mystics know through their experience that the soul is always identical with God, but is concealed from the awareness of this unity by the (veil of) ignorance inherent in phenomenal manifestation. The central teaching of all genuine religious teachers, is that the inner Self and God are one. This is expressed in the Upanishadic dictum: *tat twam asi*, "That thou art." It is this very knowledge, experienced in a moment of clarity in contemplation or prayer, which prompted Jesus of Nazareth to explain to his disciples who he was, and who they were, eternally:

> If you knew who *I* am, you would also know the Father. Knowing me, you know Him; seeing me, you see Him. ... Do you not understand that I am in the Father and the Father is in me? ... It is the Father who dwells in me doing His own work. Understand me when I say that I am in the Father and the Father is in me.[12]

There are many other Nondualist teachings, which one can find in the utterances of Jesus, and his followers. For example, it follows from the teaching of Nonduality—that is to say, the teaching that all beings are manifestations of the one Divinity, that we should therefore treat all beings as our own Self, as they most truly are. We find this teaching very prominent among the teachings of Jesus. In his Sermon on The Mount, he says:

> Ye have heard that it has been said, thou shalt love thy neighbor, and hate thine enemy; but I say unto you, love your enemies [also]; bless them that curse you, do good to them that hate you, and pray for them which despitefully use you, and persecute you; that you may be the children of your Father which is in heaven; for He maketh His sun to rise on the just and on the unjust. Be ye therefore perfect, even as your Father which is in heaven is perfect.[13]

This is the message of equality-consciousness, of seeing God (one's eternal Self) in all beings, and of thinking and acting for the benefit of all. It is this kind of reformation of our minds and hearts that is called for if we

The Divine Universe

are to assume our true identity, and experience the perfection of our eternal Self. It is, of course, our own minds which must be transformed if we are to be capable of ridding ourselves of the false notion of a separate and distinct identity apart from the one eternal Identity. It is the mind, which must be made single, one-pointed, and eventually identified with the eternal Self.

To this end, Jesus spoke to his disciples of the necessity of releasing their minds from concerns for the welfare of their separate personalities and worldly holdings in order to lift them up to God through meditation and prayer. "How," he asked them, "can you have your mind on God and at the same time have it occupied with the things of this world?" He pointed out to them that their hearts would be with that which they valued most. One's attention could not be focused on God and on one's worldly concerns at the same time, for, as he said, a city divided against itself must fall. He advised them frequently to let God be the sole focus of their attention, and to let God be the sole master whom they served. "No man can serve two masters," he said;

> for either he will hate the one, and love the other, or else he will hold to the one, and despise the other. Ye cannot serve both God and Mammon [the flesh]. Therefore, I say unto you: take no thought for your life, what ye shall eat, or what ye shall drink; nor yet for your body, what ye shall put on. For your heavenly Father knoweth that ye have need of all these things. But seek ye first the kingdom of God, and His righteousness; and all these things shall be added unto you. [14]

Naturally, this is a hard saying to those who harbor many hopes and dreams of individual worldly wealth and attainments. You'll recall what Jesus said to the sincerely spiritual man who, nonetheless, was yet attached to his worldly wealth; "It would be easier for a camel to go through the eye of a needle," he said, "than for such a man to experience the kingdom of God." The necessity for renouncing the preoccupation of the mind with worldly things if one is to occupy the mind with thoughts of God, is a teaching that is found, not only in Vedanta and Christianity, but in all true religion. It is certainly a consistently recognized fact within the long tradition of Christian mysticism. Listen, in this regard, to the words of the 5th century Christian mystic who called himself Dionysius the Areopagite:

> While God possesses all the positive attributes of the universe, yet, in a more strict sense, he does not possess them, since He transcends them all. [15]

> ... The all-perfect and unique Cause of all things transcends all, (and) is free from every limitation and beyond them all. [16] Therefore, do thou, in the diligent exercise of mystical contemplation, leave behind the senses and the operations of the intellect, and all things sensible and intellectual, and all things in the world of being and non-being, that thou mayest arise by unknowing towards the union, as far as is attainable, with Him who transcends all being and all knowledge. For by the unceasing and absolute renunciation of thyself and of all things, thou mayest be born on high, through pure and entire self-abnegation, into the superessential radiance of the Divine.[17]

We are accustomed, perhaps, to associating the word, "renunciation" with the Vedantic tradition of India, and most especially as it is used in the *Bhagavad Gita*; but renunciation of the false individual self is a prerequisite to God-consciousness, regardless of one's nationality or religious affiliation. It is a word, which occurs frequently among the writings of the great Christian mystics of the past. Listen, for example, to the 16th century Spanish monk, St. John of the Cross:

> The road and ascent to God necessarily demands a habitual effort to renounce and mortify the appetites; and the sooner this mortification is achieved, the sooner the soul reaches the summit. But until the appetites are eliminated, a person will not arrive, no matter how much virtue he practices. For he will fail to acquire perfect virtue, which lies in keeping the soul empty, naked, and purified of every appetite. [18]
> Until slumber comes to the appetites through the mortification of sensuality, and until this very sensuality is stilled in such a way that the appetites do not war against the Spirit, the soul will not walk out to genuine freedom, to the enjoyment of union with its Beloved. [19]

Now, I would like for you to hear one more Christian seer on this same theme: Thomas á Kempis was a German monk of the 15th century who, above all other mystics, Christian or Vedantic, had a great influence upon me for the beauty of his expression and the pure sincerity of his longing for God. Here is just a little of what he had to say:

> You may in no manner be satisfied with temporal goods, for you are not created to rest yourself in them. For if you alone

might have all the goods that ever were created and made, you might not therefore be happy and blessed; but your blessedness and your full felicity stands only in God who has made all things. And that is not such felicity as is commended by the foolish lovers of the world, but such as good men and women hope to have in the bliss of God, and as some spiritual persons, clean and pure in heart, sometimes do taste here in this present life, whose conversation is in heaven. All worldly solace and all man's comfort is vain and short, but that comfort is blessed and reliable that is perceived by the soul inwardly in the heart.

Await, my soul, await the promise of God, and you shall have abundance of all goodness in Him. If you inordinately covet goods present, you shall lose the Goodness eternal. Have therefore goods present in use and Goodness eternal in desire.[20]

Here, again, from the same author:

Many desire to have the gift of contemplation, but they will not use such things as are required for contemplation. And one great hindrance of contemplation is that we stand so long in outward signs and in material things, and take no heed of the perfect mortifying of our body to the Spirit. I know not how it is, nor with what spirit we are led, nor what we pretend, we who are called spiritual persons, that we take greater labor and study for transitory things than we do to know the inward state of our own soul. But, alas for sorrow, as soon as we have made a little recollection to God, we run forth to outward things and do not search our own conscience with due examination, as we should, nor heed where our affection rests, nor sorrow that our deeds are so evil and so unclean as they are. [21]

... You shall much profit in grace if you keep yourself free from all temporal cares, and it shall hinder you greatly if you set value on any temporal thing. Therefore, let nothing be in your sight high, nothing great, nothing pleasing nor acceptable to you, unless it be purely God, or of God. Think all comforts vain that come to you by any creature. He who loves God, and his own soul for God, despises all other love; for he sees well that God alone, who is eternal and incomprehensible, and fulfills all things with His goodness, is the whole solace and comfort of the soul; and that He is the very true gladness of heart, and none other but only He.[22]

This grace is a light from heaven and a spiritual gift of God. It is the proper mark and token of elect people and a guarantee of the everlasting life. It lifts a man from love of earthly things to the love of heavenly things, and makes a carnal man to be a man of God. And the more that nature is oppressed and overcome, the more grace is given, and the soul through new gracious visitations is daily shaped anew and formed more and more to the image of God.[23]

Thus, as we have seen, the true religion, the true understanding, is always the same. The teachings of the saints who have known their true nature as Divine have always declared the same path of one-pointed devotion as the means to experience and become aware of their Identity as the Divine Self. And so we find, in the words of the mystics of Christianity, Islam, Buddhism, and of every true religious tradition, the authentic teachings of Nondualism.

Part Two

Once we begin to look at the teachings of Jesus in the light of his "mystical" experience of Unity, we begin to have a much clearer perspective on all the aspects of the life and teaching of the man. His teachings, like those of the various Vedantic sages who've taught throughout the ages, is that the soul of man is none other than the one Divinity, none other than God; and that this Divine Identity can be experienced and known through the revelation that occurs inwardly, by the grace of God, to those who prepare and purify their minds and hearts to receive it. The words of Jesus are so well known to us from our childhood that, perhaps, they have lost their meaning through our over-familiarity with them. He attempted to explain to us, with the words, "I and the Father are one," that the "I," our own inner awareness of self, is none other than the one Self, the one Awareness, the Lord and Father of us all.

Why, then, are we so unable to see it? Why should it be so hard for us to attain to that purity of heart, which Jesus declared so essential to Its vision? Probably because we have not really tried—not the way Jesus did, going off into the wilderness, jeopardizing everything else in his life for this one aim, following his inspiration and focusing completely and entirely on attaining the vision of God. Not the way the Buddha did. Not the way all those who have experienced God have done. Perhaps we're not ready for such a concentrated effort just yet. Perhaps we have other desires yet to dispense with before we will be free enough to seek so high a goal. For us, perhaps, there is yet much to be done to soften the heart, so that we are pure

The Divine Universe

enough to hear the call of Divine Grace. It is to such as us, for whom much yet needs to be accomplished toward the attainment of a "pure heart," that Jesus spoke.

All of what Jesus taught to his disciples was by way of explaining to them that his real nature, and that of all men, is Divine; and that the reality of this could be realized directly. Furthermore, he taught them the path, or method, to follow in order to attain this direct realization. Let us look to his own words to corroborate this: In the Gospel book of John, he laments to God, "O righteous Father, the world has not known Thee. But I have known Thee." [24] And, as he sat among the orthodox religionists in the Jewish temple, he said, "You say that He is your God, yet you have not known Him. But I have known Him." [25] Jesus had "known" God directly during a time of deep prayer, following his initiation by his "guru," John the Baptist, probably during his time in the wilderness; and that experience had separated him and effectively isolated him from his brothers, because he alone among his contemporaries seemed to possess this rare knowledge of the Truth of all existence.

This is the difficult plight of all those who have been graced with "the vision of God." It is the greatest of gifts, it is the greatest of all possible visions; and yet, because the knowledge so received is completely contrary to what all men believe regarding God and the soul, it is a terribly alienating knowledge, which in every time has brought upon its possessor the scorn and derision of all mankind. History is replete with examples of others who, having attained this saving knowledge, found the world unwilling to accept it, and ready to defend its ignorance aggressively. This circumstance is little changed today.

Because the "vision" of God is so difficult to convey to those who had not experienced it, Jesus spoke often by way of analogy or metaphor in order to make his meaning clear. He spoke of the experience of "seeing" God as entering into a realm beyond this world, a realm where only God is. In his own Aramaic language, he called this realm *malkutha*. In the Greek translation, it is *basileia*. In English, it is usually rendered as "the kingdom of God."

> His disciples asked him, "When will the kingdom come?" Jesus said, "It will not come by waiting for it. It will not be a matter of saying 'Here it is!' or 'There it is! Rather, the kingdom of the Father is [already] spread out upon the earth, and [yet] men do not see it. [26]
>
> ... Indeed, what you look forward to has already come, but you do not recognize it." [27]

The Pharisees asked him, "When will the kingdom of God come?" He said, "You cannot tell by signs [i.e., by observations] when the kingdom of God will come. There will be no saying, "Look, here it is!" or "There it is!" For, in fact, the kingdom of God is [experienced] within you." [28]

Jesus said, "If those who lead you say to you, "See, the kingdom is in the sky," then the birds of the sky will have preceded you. If they say to you, "It is in the sea," then the fish will precede you. Rather the kingdom is inside of you, and it is outside of you [as well]. When you come to know your Self, then you [i.e., your true nature] will be known, and you will realize that it is you who are the sons of the living Father. But if you will not know your Self, you live in poverty [i.e., you live in the illusion that you are a pitiful creature far from God]." [29]

Another of Jesus' metaphors utilized the terms, "Light" and "darkness" to represent the Divinity and the inherent delusion of man, respectively:

Jesus said, "The world's images are manifest to man, but the Light in them remains concealed; within the image is the Light of the Father. He becomes manifest as the images, but, as the Light, He is concealed." [30]

He said to them, "There is a Light within a man of Light, and It lights up the whole world. If it does not shine, he is in darkness." [31]

These are terms, which have been used since time immemorial to represent the Divine Consciousness in man and the hazy ignorance, which obscures It. These two terms, "Light and "darkness," are also indicative of the cosmic aspects of Reality; in other words, they are not only the Divine Consciousness in man and the darkness of unknowing, but they are, at a higher level, the very Godhead and Its Power of manifestation. They are those same two principles we have so often run into, called "Brahman and Maya," "Purusha and Prakrti," "Shiva and Shakti." It is the Godhead in us, which provides the Light in us; it is the manifestory principle, which, in the process of creating an individual soul-mind-body, provides us with all the obscuration necessary to keep us in the dark as to our infinite and eternal Identity.

The Divine Universe

>Jesus said, "If they ask you, 'Where did you come from?' say to them, 'We came from the Light, the place where the Light came into being of Its own accord and established Itself and became manifest through our image.'
>
>"If they ask you, 'Are you It?' say, 'We are Its children, and we are the elect of the living Father.' If they ask you, 'What is the sign of your Father in you?' say to them, 'It is movement and repose.'"[32]
>
>Jesus said, "I am the Light; I am above all that is manifest. Everything came forth from me, and everything returns to me. Split a piece of wood, and I am there. Lift a stone, and you will find me there." [33]

Here, Jesus identifies with the Eternal Light; but he seems never to have intended to imply that he was uniquely and exclusively identical with It; it should be clear that his intention was always to convey the truth that all men are, in essence, the transcendent Consciousness, manifest in form:

>Ye *are* the Light of the world. Let your Light so shine before men, that they may see your good works, and glorify your Father which is in heaven. [34]

Frequently he declared to his followers that they too would come to the same realization that he had experienced:

>"I tell you this," he said to them; "there are some of those standing here who will not taste death before they have seen the kingdom of God already come in full power."[35]

>"The heavens and the earth will be rolled up in your presence. And the one who lives from the living ONE will not see death. Have I not said: 'whoever finds his Self is superior to the world?'" [36]

>"Take heed of the living ONE while you are alive, lest you die and seek to see Him and be unable to do so."[37]

>"That which you have will save you if you bring It forth from yourselves. That which you do not have within you will destroy you." [38]

"That which you have" is, of course, the Truth, the Light, the Divinity who manifests as you. "That which you do not have" refers to the false identity of separate individuality, which is simply a lie. It is the wrong understanding of who you are that limits you, causes you to identify with suffering, and prevents you from experiencing the Eternal. The teaching, common to all true "mystics" who have realized the Highest, is "You *are* the Light of the world! You *are* That! Identify with the Light, the Truth, for That is who you really are!" And yet Jesus did not wish that this should remain a mere matter of faith with his disciples; he wished them to realize this truth for themselves. And he taught them the method by which he had come to know God. Like all great seers, he knew both the means and the end, he knew both the One and the many. Thus we hear in the message of Jesus an apparent ambiguity, which is necessitated by the paradoxical nature of the Reality.

In the One, the two—soul and God—play their love-game of devotion. At one moment, the soul speaks of God, its "Father"; at another moment, it is identified with God, and speaks of "I." Likewise, in the words of Jesus to his disciples, we see this same complementarity: At one moment, he speaks of dualistic devotion in the form of prayer ("Our Father, who art in heaven"); and at another moment he asserts his oneness, his identity, with God ("Lift the stone and I am there ..."). But he cautioned his disciples against offending others with this attitude ("If they ask you, 'Are you It?' say, 'We are Its children ...'").

At times, identifying with the One, he asserts that he has the power to grant the experience of Unity ("I shall give you what no eye has seen and what no ear has heard and what no hand has touched and what has never occurred to the human mind"). [39] And at other times, identifying with the human soul, he gives all credit to God, the Father ("Why do you call me good? There is no one good but the ONE, that is God."). [40]

There is an interesting story that appears in both Matthew and Luke which illustrates the knowledge, from the standpoint of the individual soul, that the realization of God comes, not by any deed of one's own, but solely by the grace of God: Jesus had just commented upon how difficult it would be for a young man, otherwise spiritually inclined, who was attached to his worldly wealth and occupations, to realize God; and his disciples, who were gathered around, were somewhat disturbed by this, and asked, "Then, who *can* attain salvation?" And Jesus answered, "For man it is impossible; but for God, all things are possible."

And Peter, understanding that Jesus is denying that any man, by his own efforts, can bring about that experience, but only God, by His grace,

The Divine Universe

gives this enlightenment, objected: "But we here have left our belongings to become your followers!" And Jesus, wishing to assure them that any effort toward God-realization will bear its fruits in this life and in lives to come, said to them: "I tell you this; there is no one who has given up home, or wife, brothers, parents or children, for the sake of the kingdom of God, who will not be repaid many times over in this time, and in the time to come [will] know eternal Life." [41] He could guarantee to no one that knowledge of God; that was in the hands of God. But Jesus knew that whatever efforts one makes toward God must bear their fruits in this life, and in the lives to come.

And so, throughout the teachings of Jesus, one finds these two, apparently contradictory, attitudes intermingled: the attitude of the *jnani* ("I am the Light; I am above all that is manifest"); and the attitude of the *bhakta* ("Father, father, why hast Thou forsaken me?"). They are the two voices of the illumined man, for he is both, the transcendent Unity and the imaged soul; he has "seen" this unity in the "mystical experience."

Jesus had experienced the ultimate Truth; he had clearly seen and known It beyond any doubt; and he knew that the consciousness that lived as him was the one Consciousness of all. He knew that he was the living Awareness from which this entire universe is born. This was the certain, indubitable, truth; and yet Jesus found but few who could even comprehend it. For the most part, those to whom he spoke were well-meaning religionists who were incapable of accepting the profound meaning of his words. The religious orthodoxy of his time, like all such orthodoxies, fostered a self-serving lip-service to spiritual ideals, and observed all sorts of symbolic rituals, but was entirely ignorant of the fact that the ultimate Reality could be directly known by a pure and devout soul, and that this was the real purpose of all religious practice.

Jesus realized, of course, that despite the overwhelming influence of the orthodox religionists, still, in his own Judaic tradition, there had been other seers of God, who had known and taught this truth. "I come," said Jesus, "not to destroy the law [of the Prophets], but to fulfill it." [42] He knew also that any person who announced the fact that they had seen and known God would be persecuted and belittled, and regarded as an infidel and a liar. In the *Gospel of Thomas*, Jesus is reported to have said, "He who knows the Father (the transcendent Absolute) and the Mother (the creative Principle) will be called a son-of-a-bitch!" [43] It seems he was making a pun on the fact that one who does *not* know his father and mother is usually referred to in this fashion; but, in his case, he had known the Father of the universe, and knew the Power (of Mother Nature) behind the entire creation, and still he was called this derisive name. Such derision is the common experience of all the great seers, from Lao Tze to Socrates and Heraclitus, from Plotinus

The Perennial Philosophy

and al-Hallaj to Meister Eckhart and St. John of the Cross. All were cruelly tortured and persecuted for their goodness and wisdom. Jesus too found the world of men wanting in understanding; he said:

> I took my place in the midst of the world, and I went among the people. I found all of them intoxicated [with pride and ignorance]; I found none of them thirsty [for Truth]. And my soul became sorrowful for the sons of men, because they are blind in their hearts and do not have vision. Empty they came into the world, and empty they wish to leave the world. But, for the moment, they are intoxicated; when they shake off their wine, then they will repent. [44]

Notes:

1. This Essay is reprinted, in a slightly revised form, from my earlier work, *The Wisdom of Vedanta*, Olympia, Wash., Atma Books, 1991, 1994; reissued by O Books in 2006.

2. Meister Eckhart, *Sermon 6*; Colledge, E. & McGinn, B. (trans.), *Meister Eckhart: The Essential Sermons, etc.;* p.188.

3. Meister Eckhart, *Sermon 18;* Blackney, R.B., *Meister Eckhart: A Modern Translation;* P. 181.

4. Nicholas of Cusa, *De Visio Dei, XXV;* Salter, E.G., *The Vision of God;* p. 129.

5. *Ibid., XVI;* p. 78.

6. *Ibid., XII;* p. 56.

7. *Ibid., XVII;* pp. 81-82.

8. *Ibid., XIV;* p. 66.

9. John of the Cross, *Spiritual Canticle, 26:4;* Kavanaugh, K. & Rodriguez, O. (trans.), *The Collected Works of John Of The Cross;* p. 512.

10. *Ibid., 22:3-4;* p. 512.

11. *Ibid., Living Flame Of Love, III:78;* p. 641.

12. *Gospel of John, 13:40.*

13. *Gospel of Matthew, 5:43-48*

14. *Ibid., 6:24-25, 31-33.*

15. Pseudo-Dionysius, *Mystical Theology, I;* Editors Of The Shrine Of Wisdom, *Mystical Theology, etc.;* p. 10.

16. *Ibid., V;* p. 16.

17. *Ibid., I;* p. 10.

18. John of the Cross, *The Ascent of Mount Carmel, I:5:6;* Kavanaugh & Rodriguez, *Op. Cit.;* p. 83.

19. *Ibid., I:15:2;* p. 106.

20. Thomas á Kempis, *De Imitatio Christi, III:9;* Abhayananda, S., *Thomas á Kempis: On The Love of God;* pp. 109-110.

21. *Ibid., III:3;* p. 90.

22. *Ibid., II:5;* p. 70.

23. *Ibid., III:4;* pp. 95-96.

24. *Gospel of John, 17:25.*

25. *Ibid., 8:54.*

26. *Gospel of Thomas, 114;* Robinson, James M. (trans.), *The Nag Hammadi Library;* pp. 118-130.

27. *Ibid., 51.*

28. *Gospel of Luke, 17:20.*

29. *Gospel of Thomas, 3;* Robinson, J.M., *Op. Cit..*

30. *Ibid., 83*

31. *Ibid., 24.*

32. *Gospel of Thomas,* 50.

33. *Ibid., 77.*

34. *Gospel of Matthew,* 5:14-16.

35. *Gospel of Mark,* 9:1.

36. *Gospel of Thomas,* 111; Robinson, J.M., *Op. Cit.*

37. *Ibid.* 59.

38. *Ibid.,* 70.

39. *Ibid.,* 17.

40. *Gospel of Luke,* 18:18.

41. *Ibid.,* 18:18-30; *Matthew,* 19:16.

42. *Gospel of Matthew,* 5:17.

43. *Gospel of Thomas,* 105.

44. *Ibid.,* 28.

21.

THE MEETING OF HEART AND MIND [1]

There is a saying that the man of devotion (the *bhakta*) and the man of knowledge (the *jnani*) are like a blind man and a lame man, respectively. Neither can get about on his own; the *bhakta* without discrimination isn't able to see where he's going, and the *jnani* without heart is lame and unable to go forward. A happy solution is found to both their problems, however, when the lame *jnani* is hoisted upon the shoulders of the blind *bhakta*. For then, the *jnani* provides the *bhakta* with vision owing to his heightened position, and the *bhakta* provides the *jnani* with the means of locomotion. The point of this saying, of course, is that this is what we must do with the two sides of our own nature: we must combine them and utilize both, so we have the benefit of both discriminative knowledge and the sweetness of devotion.

In the spiritual life, the intellect and the heart play equally important parts. Like the blind man and the lame man, each is helpless without the other. Just think: how many times do we meet up with a simple, good-hearted person, full of sincere love for God, and yet who, because of a lack of discrimination, becomes lost on a path which leads only to a gushy sentimentality and misplaced affections. And how often also do we see the overly intellectual, the stiff, proud person unwilling to let go of concepts long enough to feel the joy of love, or to simply pray with a humble, contrite, and loving heart.

Clearly, both are equally handicapped. The heart without discrimination leads one only into darkness and confusion. And the intellect without the sweetness of the heart makes of life a dry and trackless desert, without any flavor or joy. It is my considered opinion that if a person is to reach the highest perfection possible to man, there must be a balance of heart and mind. There must be both the knowledge of the Self, and at the same time, the love of God.

One of the greatest works of Spiritual devotion, the *Srimad Bhagavatam*, states: "The essence of all one's efforts should consist in withdrawing the mind from the objects of sense, and fixing it on God alone." Continuing, it says, "The mind must be engaged in one thing or another: if it meditates on sense-objects, it becomes worldly; if it meditates on God, it becomes Divine."

All the great scriptures similarly extol in one way or another the focusing of the mind on God. Some call it "devotion"; some call it "awareness of the Self." Narada, who was the epitome of the *bhakta*, states in his *Bhakti Sutras*, "The constant flow of love towards the Lord, without any selfish desire, is devotion." And Shankaracharya, who was the *jnani* of *jnanis*, says, "Devotion is continuous meditation on one's true Self." Now, if we examine the matter closely, we can see that devotion to God is not in any way different from meditation on the Self; and that the experience of Divine Love is not different from the experience of the Bliss of the Self.

The mind experiences Unity as Consciousness and Bliss. The heart experiences God as the fullness of Love and Joy. Are 'Bliss' and 'Love' different in any way? If the heart sings of God, does that take anything away from His Bliss? If the mind is aware of the Self, does that take anything away from His Love? The Truth remains, whether we make a noise or keep silent; whether we give Him this name or that, He remains the same. Whether we regard ourselves as the worshipper or the worshipped, there is nothing here but the One. Whether we call our intrinsic happiness by the name of Bliss or Love, its taste remains the same. We may call Him whatever name we like; we may sing it out to our heart's content. Whether we are gamboling in the streets or sitting quietly at home, we are always God playing within God. To remember Him is our only happiness; to forget Him our only sorrow.

When we speak of Self-knowledge, we must differentiate between such Knowledge as is identical with the Bliss of the Self and that knowledge which is simply the knowledge of such Knowledge. Intellectual knowledge of Unity (Nondualism) is a wonderful thing, but it is only preparatory to true Knowledge, that Knowledge which is synonymous with true enlightenment. Conceptual knowledge we must certainly go beyond. To do so, it is necessary to utilize the heart. Devotion leads the mind beyond mere intellectual knowledge to the experience of the Blissful Self—which is true Knowledge.

The 19th century saint, Sri Ramakrishna, was fond of bringing out this truth in his conversations and his songs. Here is one such song:

> How are you trying, O my mind, to know the nature of God?
> You are groping like a madman locked in a dark room.
> He is grasped through ecstatic love;
> How can you fathom Him without it?
> And, for that love, the mighty yogis
> Practice yoga from age to age.
> Then, when love awakes, the Lord, like a magnet,
> Draws to Him the soul.
> It is in love's elixir only that He delights, O mind!

The Divine Universe

He dwells in the body's inmost depths, in everlasting Joy.

Sri Ramakrishna himself became so full of desire for God, whom he regarded as his "Mother," that people began to fear for his sanity when they would see him rubbing his face on the ground and weeping for his "Mother" to come. At times, he would sing this song:

> O Mother, make me mad with Thy love!
> What need have I of knowledge or reason?
> Make me drunk with Thy love's wine!
> O Thou, who stealest Thy bhakta's hearts,
> Drown me deep in the sea of Thy love!
> Here in this world, this madhouse of Thine,
> Some laugh, some weep, some dance for joy:
> Jesus, Buddha, Moses, Gauranga—
> All are drunk with the wine of Thy love.
> O Mother, when shall I be blessed
> By joining their blissful company?

Such total abandon, such complete disregard for one's own reputation, status, future welfare, is typical of those who, in the end, attain to God. The great poet-saint, Kabir, spoke often of the need to renounce all other desires in order to attain God. "Love based on desire for gain," he said, "is valueless! God is desireless. How then, could one with desire attain the Desireless?" Kabir then went on to say, "When I was conscious of individual existence, the love of God was absent in me. When the love of God filled my heart, my lesser self was displaced. O Kabir, this path is too narrow for two to travel."

You see, in the experience of the One, there's no place for two; one of the two must go. Whether your focus is on God or on the Self, you must transcend the (illusory) separate self, the ego. The path of love, says Kabir, is too narrow for two to travel; the ego must yield to the Beloved. "Very subtle," he says, "is the path of love! There, one loses one's self at His feet. There, one is immersed in the joy of the seeking, plunged in the depths of love as the fish in the depths of the water. The lover is never slow in offering his head for his Lord's service. This, Kabir is declaring, is the secret of love."

"How odd!" you may think; "Must I really offer my life, be willing to give up my head in order to attain God?" Let me tell you a story: it is a story from the *Masnavi*, the Persian masterpiece of the great Sufi poet, Jalaluddin Rumi. In it, he tells the story of the Vakil of Bukhara. The Vakil is the prince; he represents the supreme Lord. One of the subjects of this prince is told that the Vakil is seeking him for the purpose of chopping off his head. The poor man, hearing this, flees the city into the desert, and wanders from small village to village, in his attempt to stay out of the hands of the Vakil.

For ten years the man runs and runs. Then, finally exhausted and humiliated, he returns in surrender to Bukhara. The people there who knew him previously shout to him from their homes: "Escape while you can! Run! Run for your life!" But the man continues to walk in the direction of the Vakil's palace. "The Vakil is searching everywhere for you," they cry; "He has vowed to cut off your head with his own sword!" And, while everyone was shouting their warnings to this man, he just kept walking toward the palace of the prince. The people were calling to him from right and left: "Are you mad?" they shouted; "You are walking into certain death! Run! Run, while you have the chance!" But the man kept on walking, right into the palace of the Vakil.

When he reached the Vakil's antechamber, he entered it and walked right up to the throne, then he threw himself on the floor at the prince's feet. "I tried to escape you," the man said, "but it is useless. My heart knows that my greatest destiny is to be slain by you. Therefore, here I am; do with me what you will." But, of course, the prince had no desire to slay the man; he was very pleased, though, to see that the man had surrendered to him even when he thought he would lose his head thereby. And so the Vakil raised the man up and made him his representative throughout the realm. And Rumi, the author of this story, says at the end, "O lover, cold-hearted and unfaithful, who out of fear for your life shun the Beloved! O base one, behold a hundred thousand souls dancing toward the deadly sword of his love!"

This is a recurrent theme among the devotional poets of the Sufi tradition. Kabir, whom I quoted a moment ago, asks of the devotee:

> Are you ready to cut off your head and place your foot on it?
> If so, come; love awaits you!
> Love is not grown in a garden, nor sold in the marketplace.
> Whether you are a king or a servant,
> The price is your head and nothing else.
> The payment for the cup of love is your head!
> O miser, do you flinch? It is *cheap* at that price!
> Give up all expectation of gain.
> Be like one who has died, alive only to the service of God. Then God will run after you, crying, "Wait!
> Wait! I'm coming."

It is clear, of course, that what is necessary is not one's physical death, but the death of the ego-self. The little identity of "me" and "mine" is to be sublimated into the greater Identity of the one all-pervading Self through a continuous offering of the separative will into the universal will, an offering of the separative mind into the universal Mind, and the offering of the individual self in service of the universal Self.

Sri Ramakrishna knew very well how persistent is this false sense of ego, of selfhood. For this reason, he taught, not the suppression of this ego, such as the *jnani* practices, but rather the utilization of the ego in devotion and service to God. "The devotee," says Sri Ramakrishna, "feels, 'O God, Thou art the Lord and I am Thy servant.' This is the 'ego of devotion.' Why does such a lover of God retain the 'ego of devotion?' There is a reason. The ego cannot be gotten rid of; so let the rascal remain as the servant of God, the devotee of God."

You see, Sri Ramakrishna understood that, so long as this universe exists, the apparent duality of soul and God exists. Until such time as God merges the soul into Himself, both of these exist. We are the absolute Consciousness, to be sure; but we are also His manifested images. We are Brahman, but we are also (part of) Maya; we are Shiva, but we are also Shakti; we are the universal Self, but we are also the individualized self. It is foolish not to acknowledge both sides of our nature. Failing to do so only leads us into great conflicts and difficulties. If we deny and neglect the existence of the soul, asserting only, "I am the one pure Consciousness," the active soul will rise up and make us acknowledge its presence. The only way to lead the soul to the experience of its all-pervasiveness is to teach it love for God, to transform it into Divine Love. The soul that goes on expanding its power to love eventually merges into absolute Love, and awakes to the truth that it *is* Love.

Remember, whatever you continually think of for a long time, that you become. So, if the mind continually thinks of God, it will attain the state of Love. No amount of knowledge will awaken the mind to love. Nor will the mind become quieted by force or the power of will. It will only become more frustrated, agitated and antagonistic. Instead of trying to do violence to the mind, lead it into meditation by the path of love. Soak it in the vat of love, and dye it in the crimson color of love; then it will merge into the sweetness of God.

I'd like to share with you a few words of inspiration from a modern saint who extolled this very truth of devotion to God for many years. In my search for someone who best represented the synthesis of the heart and mind, I considered many different saints, both ancient and modern. But, it seemed to me that one of the very best examples that could possibly be held up is that of a woman who was called Anandamayee Ma, "the Bliss-permeated Mother." Anandamayee Ma (1896-1982) [2] is mentioned in Yogananda's *Autobiography Of A Yogi*, as a saint whom he met in 1935. Even then, she was a remarkable woman, inspiring everyone with whom she came in contact by her simple purity, and the depth of her God-realization.

She was born in 1896 in what is now Bangladesh. Since the mid-1920's she has been one of the most revered saints in all of India. She stayed in one place for only brief periods, preferring to travel about India, visiting her many devotees here and there. She passed from life in 1982, leaving this world a poorer place. For she was the epitome of a *jnani*, with the heart of a *bhakta*. Her exposition of the Self, from the standpoint of the philosophy of Nondualism, was flawless. She possessed the shining intellect of a god. She was always poised in the highest state. And yet, she was also a humble servant of God, exhorting others to give all their devotion to God alone. Listen to what she had to say:

> It is by crying and pining for Him that the One is found. In times of adversity and distress as well as in times of well being and good fortune, try to seek refuge in the One alone. Keep in mind that whatever He, the All-Beneficent, the Fountain of Goodness, does, is wholly for the best. He alone knows to whom He will reveal Himself and under which form. By what path and in what manner He attracts any particular person to Himself is incomprehensible to the human intelligence. The path differs for different pilgrims. The love of God is the only thing desirable for a human being. He who has brought you forth, He who is your father, mother, friend, beloved and Lord, who has given you everything, has nourished you with the ambrosia streaming from His own being—by whatever name you invoke Him, that name you should bear in mind at all times.

> Apart from seeking refuge in the contemplation of God, there is no way of becoming liberated from worldly anxiety and annoyance. Engage in whatever practice that helps to keep the mind centered in Him. To regret one's bad luck only troubles the mind and ruins the body; it has no other effect—keep this in mind! He by whose law everything has been wrought, He alone should be remembered. Live for the revelation of the Self hidden within you. He who does not live thus is committing suicide. Try to remove the veil of ignorance by the contemplation of God. Endeavor to tread the path of immortality; become a follower of the Immortal.

... Meditate on Him alone, on the Fountain of Goodness. Pray to Him; depend on Him. Try to give more time to *japa* (repeating His name) and meditation. Surrender your mind at His feet. Endeavor to sustain your *japa* and meditation without a break.

It is necessary to dedicate to the Supreme every single action of one's daily life. From the moment one awakes in the morning until one falls asleep at night, one should endeavor to sustain this attitude of mind. ... Then, when one has sacrificed at His feet whatever small power one possesses, so that there is nothing left that one may call one's own, do you know what He does at that fortunate moment? Out of your littleness He makes you perfect, whole, and then nothing remains to be desired or achieved. The moment your self-dedication becomes complete, at that very instant occurs the revelation of the indivisible, unbroken Perfection, which is ever revealed as the Self.

These words of Anandamayee Ma constitute the ancient, yet ever-new, message of all the saints. Knowledge is essential to clear away our doubts, to understand where our greatest good lies. But it is devotion that takes us to our Destination. The determined dedication of the heart, mind, and will to God is the means to fulfillment, and the means to the perfect Knowledge which is the Self.

Notes:

1. This Essay is excerpted from my previously published book, *The Wisdom of Vedanta*, Olympia, Wash., Atma Books, 1994; it does not appear in the subsequently published edition by O Books, London, 2005.

2. For details on the life and teaching of Anandamayee Ma, see the excellent book by Timothy Conway, Ph.D., *Women of Power and Grace: Nine Astonishing, Inspiring Luminaries of Our Time*, The Wake Up Press, 1994.

IV.

THE SCIENCE OF THE SOUL

We have established that the eternal Self, or God, is the ultimate Identity of everyone; but what is it that makes "individuals" of each of us? Why does one person have a passion for music and another for physics? Why does this one become an architect, and the other becomes a writer of fiction; this one a neurobiologist and this other a stock broker? These are questions about the individual soul-characteristics with which each of us is born. We cannot deny that such differences exist among human beings, even though we are united as one at the Source. One person is born with every advantage; another is born infirm and in very limited circumstances. How do we account for such differences? Clearly, each individual is endowed with his or her own unique characteristics, qualities, virtues and vices by which each is set apart from the others. These may extend to physical characteristics, but they belong primarily to an interior reality, which we have traditionally termed 'the soul'.

In the Eastern religious traditions, and, in fact, in nearly every spiritual worldview, it is acknowledged that there are at least three different "levels" of subtlety leading up to the outer "gross" body of matter: Starting at the beginning, the pure Consciousness of the Absolute is the super-causal level; then it is the diffuse Energy of the Creative Power at the causal level; and from that arises an immense cosmos, and eventually each human life, containing both a subtle body formed of consciousness and a physical body formed of matter. The subtle body is what is generally regarded as "the soul", or jiva. *In one of many different versions of the constituency of the* jiva, *it is described as the repository of the* prana *(the subtle breath), the* manas, *or lower mind, from which arises the* ahamkara *(the sense of ego, or individuality), and the* buddhi, *or discriminitive intellect—all of which are colored to a great degree by our* karma *(the tendencies accumulated from the actions performed in previous incarnations).*

This soul, or subtle body, is also referred to as the "astral" body by those who assert that the soul-differences that constitute our personal uniqueness are clearly symbolized in and synchronized with the stars (astra), *or more specifically, the positions and relationships of the planets in the solar system in existence at the time of our birth. This study of the correlation of soul-qualities and planetary positions is known as* astrology; *it is sometimes referred to as "the science of the soul". Such a synchronous relationship between the Sun, moon, and planets and the human psyche, or soul, may readily appear to be an implausible if not impossible relationship if we regard it as occurring in a classic (Newtonian)*

The Divine Universe

mechanical universe. But this interconnecting relationship between the planetary environment and the soul appears in an entirely new light when it is seen to operate in a universe of Spirit, in a universe imaged in the Mind of God. And though astrology has been continually practiced for more than 3000 years, it has been, and probably shall always continue to be, practiced and understood by only a very small group of people. This is because the development of the intuitive faculty required for its comprehension and practice is confined to but an advanced few.

In this last grouping of Essays, then, I attempt to explain the value of an astrological perspective on the individual soul as an adjunct to the knowledge of the one Divine Self. And I shall start off with an account of my own rather dramatic introduction to astrological principles in connection with the event of my "mystical experience".

22.

THE ASTROLOGY OF ENLIGHTENMENT [1]

Back in the days of the hippies—1966 to be exact, I was a twenty-eight year-old recent convert to Nondual Vedanta living in Los Gatos, California. Inspired by my reading of the lives of the saints, I became a 'renunciant', and headed for the forested mountains of Santa Cruz with the aim of realizing God. At the time, I was not yet acquainted with the principles of astrology, but had I been, I would have discovered that there were a number of planetary aspects on the verge of culminating at that time which were advantageous to my avowed pursuit. My good fortune led me to an abandoned cabin in the woods with a pristine brook running by it, where I took up residence, and practiced my *sadhana* (spiritual search) for the next five years. The upshot of this story is that, on the night of November 18, 1966, I experienced what many have called a "mystical union with God". It was a deeply profound experience which utterly transformed my subsequent life, led me to India and to the vows of *sannyasa*, and the receipt of my present name. Those interested in a more detailed account of this story may find it in my book, *The Supreme Self* [2].

It was not until nearly ten years after my experience of enlightenment in the Santa Cruz mountains, that I began to be interested in the peculiar claims of astrology, and came to learn of the meaningful connections between my own natal planetary positions and my personal characteristics; and as I eagerly consumed what literature I found on the subject, I became more and more convinced of the validity of the astrological principle of correspondence between the planetary positions and the varying conditions of the human psyche. One day, it occurred to me that, if these principles were true, there would have to have been a configuration in the progressed and transiting positions of the planets on the night of my "mystical experience" ten years previous that was significantly extraordinary. In other words, the potential for that mystical experience must have been clearly signified in the planetary patterns in effect for me on that very night. That experience of union, or Unity, had come to me only once. Why on that day, at that time? I could only explain it, as the Christian saint, Thomas á Kempis did, as God's inexplicable grace. But my budding understanding of the principles of astrological correspondence had piqued my curiosity and whetted my appetite to know more. And so I drew up a chart for that night of November 18, 1966.

The Divine Universe

What a revelation it was when I beheld that chart! The correspondence was undeniable. Here before my eyes was clear and unequivocal proof of the "science" of astral correspondences. Any impartial astrologer viewing the progressions and transits to my natal chart which occurred on that evening would have to acknowledge that this was indeed a night of destiny, an undeniably magical night of mystical vision, a once-in-a-lifetime night of incredible potential for the meeting with God. The extraordinary emphasis on the planetary position of Neptune (known as the planet of mystical experience) at that particular time is eloquently conclusive.

If—as many people think—there is really no correlation between the planets and the human psyche, then what an extraordinarily grand coincidence it was, what a marvelous accident of nature, that at the same moment that I was experiencing the Godhead, the planets were proclaiming it in the heavens! I think any reasonable person with even a little astrological acumen, on viewing the "influences" in effect for me that night, would have to acknowledge that the significant planetary picture at the time of my 'enlightenment' experience does, in fact, seem to provide evidence of the validity of the contents of that experience, confirming that all things do indeed "move together of one accord," that the universe is without a doubt one interconnected and coordinated Whole.

Here is the first of two charts depicting the planetary array at the time of my 'mystical' experience. This chart, chart A, shows the *transiting* planetary arrangement in effect at the time of my "experience of unity." The lines connecting those planets in *opposition* (180°), *trine* (120°), and *sextile* (60°) aspects to each other show the angular relationships between these transiting planets. This, in itself, is a remarkable configuration. But to fully appreciate the significance of this transiting planetary arrangement, it must be seen in relationship to the positions of the planets at my birth. This may be seen in chart B, a composite chart, showing the positions of the planets in my natal, progressed, and transiting charts, in consecutive wheels. In the center wheel, my *natal* chart, calculated for 6:01 P.M., August 14, 1938, at Indianapolis, Indiana; in the intermediate wheel, my *progressed* chart for 9:00 P.M., November 18, 1966, at Santa Cruz, California; and in the outer wheel, the *transiting* chart for the same time and place.

CHART A

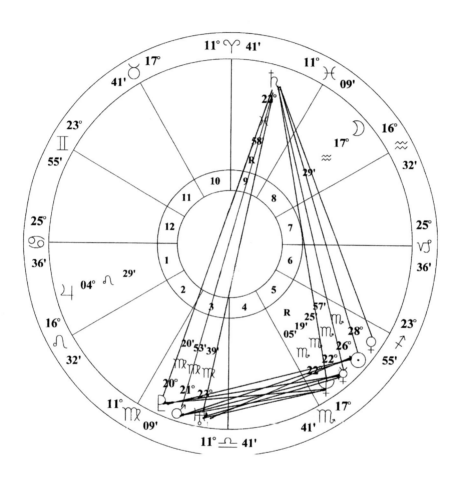

The Divine Universe

CHART B

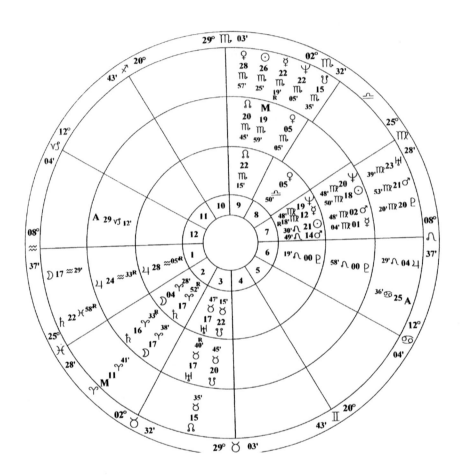

Natal Aspects:
    Sun conjunct Mars
    Sun trine Saturn
    Sun square Uranus
    Mercury trine Uranus
    Mercury conjunct Neptune
    Mars square Uranus
    Saturn semisextile Uranus
    Uranus trine Neptune

Progressed Aspects (to natal planets):
    Moon conjunct Saturn (exact)

Moon semisextile Uranus (exact)
Sun conjunct Neptune (exact)

Transiting Aspects (to natal planets):
Moon sextile Saturn (exact)
Moon square Uranus (exact)
Sun conjunct Midheaven
Mercury square Sun (exact)
Mercury conjunct North Node (exact)
Venus conjunct Midheaven (exact)
Venus square Jupiter (exact)
Mars conjunct Neptune
Jupiter trine Moon (exact)
Uranus conjunct Neptune
Neptune conjunct North Node (exact)
Neptune square Sun (exact)
Pluto conjunct Neptune (exact)

Note: planets within 1° aspect are considered to be exact.

In examining this composite of charts, the first thing that stands out to the trained eye is the highly significant progression of both the Sun and the Moon (middle wheel) to exact conjunctions with natal planets (center wheel). The Moon's progression to an exact conjunction to my natal Saturn is a conjunction which occurs only once every twenty-eight to thirty years; while the Sun's progression to the natal position of Neptune occurs in one's chart only if one's Sun position is natally within 80° or so, clockwise, of Neptune's position—and then, only once in a lifetime. The likelihood of both the Sun and Moon forming progressed conjunctions to the natal position of outer planets simultaneously is obviously very remote, and when it *does* occur, is highly significant of an extraordinary event.

Neptune, to which the progressed Sun is conjoined, figures quite prominently in my natal chart, as it forms there a conjunction to Mercury and a trine to Uranus. In my early deliberations about my own chart, I had come to look on it as a representation of a certain mental receptivity to poetic inspiration. But Neptune represents much more than that; with beneficial aspects from other planets it can represent an access to the very subtlest of spiritual realms. One astrologer, Robert Hand, who is a recognized authority on astrological symbols, says about Neptune:

The Divine Universe

> *Neptune symbolizes the truth and divinity perceived by mystics. (Keep in mind that the planet is an agent or a representation of an energy, not the source of the energy.)* At the highest level, Neptune represents Nirvana, where all individuality is merged into an infinite oneness of being and consciousness. [3]

Notice that the massive conjunction of transiting Mars-Uranus-Pluto (outer wheel) is precisely over my natal Neptune, along with the progressed Sun; and that the (exact) conjunction of transiting Mercury-Neptune is exactly conjunct my natal North Node, and trining transiting Saturn. There were, on that night of November 18, 1966, two exact conjunctions of *progressed* planets to natal planets, and ten exact aspects of *transiting* planets to natal positions, five of which were conjunctions. The concentration of energy over my natal Neptune position was clearly intense—intense enough for even a thick-headed person like myself to catch a glimpse of God.

Now, if it could be shown that, in all cases, the mystical experience of Unity coincided with progressed solar and/or lunar aspects to Neptune in the charts of the experiencers, we would be in possession of a neatly consistent formula for anticipating mystical experience. However, that does not seem always to be the case. When one examines the charts of known mystics of the past progressed to the date of their transcendent experience, one encounters a very inconsistent collection of varied influences, although aspects to the natal Neptune position do seem to figure strongly.

For example, in the chart of Sri Aurobindo (born August 15, 1872), at the time of his reported enlightenment (January 15, 1908) the progressed moon is exactly conjunct his natal Neptune, and the progressed Sun is exactly quincunx Neptune's position. In the chart of Sri Ramakrishna (born February 18, 1836), progressed to the date of his first *samadhi* at the age of twenty-nine (February 1, 1865), the progressed moon is exactly sextile his natal Neptune's position, while there are no major aspects from the progressed Sun. And in the progressed chart of Sri Ramana Maharshi (born December 30, 1879), who became enlightened at the age of sixteen (September 15, 1896), the progressed moon is 3° past a conjunction with natal Jupiter, and the progressed Sun makes only one aspect: a trine to natal Pluto. Even with so brief a sampling, it is clear that there is a wide range of variation in the progressed solar and lunar aspects occurring at the time of enlightenment.

Strangely enough, the one modern mystic whose progressed aspects at the time of his enlightenment most closely resemble the planetary aspects present in my own enlightenment chart is someone who was personally known to me and with whom I had long been associated: Swami Muktananda. Muktananda's natal horoscope reveals him to have been an immensely

The Science of the Soul

powerful personality, but it only hints at the tremendous personal power he came to possess through the legacy of *shaktipat* transmitted to him by his guru, Nityananda, and through his lifelong retention of that power. He was totally unique in his masterful attainment, and his life of sharing his spiritual realizations was also amazing and unique; but *his experience of the Self was the common experience of all the enlightened.*

While our paths to enlightenment, our visions, our circumstances, personalities and destinies (as symbolized in our individual horoscopes) were very different, the enlightenment experience which revealed the eternal Self to Muktananda was identical (by definition) with that which I experienced. What's more, the planetary significators of enlightenment were nearly identical in both our cases. Also, despite the unique elements of Muktananda's *sadhana* (spiritual search), which differed considerably from my own experience, both of our actual enlightenment experiences, though nearly ten years apart, coincided with a strong aspect of the progressed moon to one of the outer planets in the natal chart, at the same time that *the progressed Sun was forming an exact conjunction with the natal position of Neptune.* There was also, at the time, an extraordinary and significant array of transiting planets in the heavens in both cases.

The Divine Universe

Here is a chart (C) showing the positions of the transiting planets on the day of Muktananda's enlightenment (July 30, 1957):

Chart C

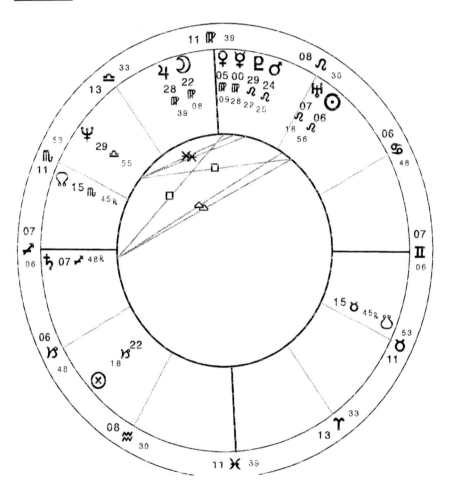

The Science of the Soul

And here is a composite chart (D) for the time of Muktananda's enlightenment (the inner wheel is his natal chart (May 16, 1908, at Mangalore, India; 6:00 AM INT); the middle wheel is the progressed chart at Yeola, India; and the outer wheel represents the transiting positions of the planets on that day, July 30, 1957):

Chart D

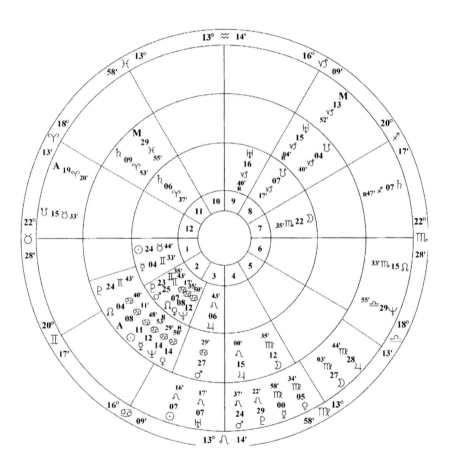

The Divine Universe

Natal Aspects:

> Sun conjunct Ascendant
> Sun opposite moon
> Mercury sextile Jupiter
> Venus conjunct Neptune
> Venus square Saturn
> Mars conjunct Pluto
> Jupiter trine Saturn (exact)
> Uranus opposite Neptune

Progressed Aspects (to natal planets):

> Sun-Mercury conjunct Neptune (exact)
> Moon sextile Neptune (exact)
> Jupiter quincunx Uranus

Transiting Aspects (to natal planets):

> Sun-Uranus conjunct Jupiter
> Sun-Uranus trine Saturn
> Moon-Jupiter trine Sun (moon exact)
> Mars-Pluto square Sun
> Saturn trine Jupiter
> Saturn trine Saturn

Note: Planets within 1° aspect are considered to be exact.

In Muktananda's *natal* chart, notice the powerful stellium of planets in the 2nd House, along with the Sun-moon opposition closely conjunct the 1st-7th House cusps. The Sun conjunct his Ascendant, Mars conjunct Pluto, and Venus conjunct Neptune give some indication of the great forcefulness of his personal energy and his spiritual evolution. Jupiter in the 3rd House shows his learning and speaking ability, and Uranus on the 9th House cusp relates both to his advanced philosophical views and his amazingly broad travels. The *progressed* chart shows the progressed Sun and Mercury in exact conjunction with his natal Neptune, and the progressed Moon in exact sextile to natal Neptune.

The transiting aspects are equally notable: a transiting Sun-Uranus conjunction is conjunct natal Jupiter, trining natal Saturn, while transiting Saturn is forming a grand trine with natal Saturn and Jupiter. A transiting Moon-Jupiter conjunction is trining the natal Sun, while transiting Mars, Pluto, and Mercury are in close square to that natal Sun position. All in all, it is a remarkable set of circumstances, signaling a remarkable occurrence. Clearly, it is *as* uniquely powerful a set of progressed and transiting aspects as those which occurred in relation to my own chart in November of 1966.

The progressed and transiting aspects shown in the charts of Swami Muktananda and myself are reminiscent of the aspects one may find in any of those other sets of charts which correspond to significant events in the lives of any other subjects—except that *these* charts signal events which have been traditionally regarded as beyond the realm of temporal occurrence. They are experiences of a transcendent reality which are regarded by many as completely unrelated to temporal 'influences' or to causal agents other than the transcendent Self, and are usually attributed to "Divine grace". The linking of "mystical experience" to corresponding planetary arrangements brings up very forcefully a number of questions: 'Was this "mystical experience" inscribed in the heavens since the beginning of time, and therefore entirely predestined? Or was it merely a *potential* opportunity? What does the suggestion of corresponding planetary factors do to the concept of grace and free will? Does it imply a karmic reservoir of previous self-effort in spiritual endeavor? And where is the hope or possibility of "spiritual experience" for those in whose astrological forecast the prerequisite planetary conditions are *not* present?' These are questions which cannot easily be answered, and which must be reflected upon by each individual. Here are the conclusions to which I have come after much deliberation:

The evolution of the soul occurs over many lifetimes, with its breakthrough summit coming with full surrender of the self in the Love of God, resulting in the subsequent realization of its transcendent and eternal Identity. And because the evolution of the universe reflects the evolution of each soul, the stellar and planetary positions, which signal that soul's enlightenment, will coincide perfectly with that moment in the soul's evolutionary summit. The question of whether it is the soul's evolutionary struggle or the planetary alignments which brings about enlightenment must be answered, "Neither." They are coordinated events in the unfolding of God's cosmic drama; both events are simultaneous effects of the one Cause, occurring in Himself in the ordered unfoldment of His will. All is one coordinated whole, and all that occurs within it is a manifestation of His grace.

The complexity of such a universe—a universe in which the evolution of each soul on earth is in synchronization with the ongoing motions of

The Divine Universe

planetary bodies—is indeed mind-boggling and well beyond our present ability to conceive or visualize. Nonetheless, we must acknowledge that it is impossible to separate the birth of any individual from the cosmic conditions in which it occurs. For the universe is an integral Whole, and every event in it is in interlocking agreement with every other; with not even the tiniest, most seemingly insignificant, event to be considered as an isolated phenomenon.

Within this Whole, where "all things move together of one accord," the division of small-scale events into categories of cause and effect is imaginary and has no real meaning. For it is the One, the Lord, God—call Him what you will—who, by means of His Power of Will, is the sole Cause of the entire manifested array of the cosmos and therefore of every single event which takes place within it. This truth is seen clearly and unmistakably in the unitive experience of the mystic.

It is my opinion that the implications and ramifications of this discovery of the correlation of planetary configurations with the inception of Divine illumination will eventually bring a revolutionary breakthrough in spiritual understanding comparable to the revolution in scientific understanding brought about by Copernicus and Galileo. However, it will require so bold a departure from traditional ways of thinking that it is unlikely to have an immediate influence on the understanding of any but the most discerning. In fact, many so-called "spiritual teachers" will find this information unacceptable and will reject it, for it negates their contention that Self-realization is solely the result of 'spiritual' practices and techniques. On the contrary, evidence shows that without the timing of the appropriate heavenly motions unfolding in one's life, no illumination will come.

Nonetheless, we must admit that the present-day understanding of how astrology 'works' is as far from a comprehensive resolution as is the science of microphysics. It was a mystery to the ancients, and it is a mystery today (although the idea of an immediate interconnectedness of everything within "the unbroken Whole" hints at the way ahead). And while the science of the 'astrology of enlightenment' is in its infancy today, I am hopeful that the data that is here provided will point the way to greater exploration and understanding of the relation of astronomical phenomena to mystical experience in the years to come. I welcome the participation and submission of pertinent data by all those with knowledge relevant to this pursuit.

Notes:
1. This Essay is excerpted from my book, *The Supreme Self,* Olympia, Wash., Atma Books, 1998; subsequently published by O Books, London, 2005.

2. *Ibid.*
3. Robert Hand, *Astrological Symbols,* Para Research, Inc., Rockford, Mass., 1980; p. 75.

23.

THE RATIONALE FOR ASTROLOGY

To the minds of some, astrology is the epitome of ignorance. Those who feel this way cringe at the very suggestion of a connection between the planetary environment of the solar system and those living within that environment, citing the absense of the slightest empirical evidence for such a claim. For many, however, the claims of astrology begin to seem plausible when standard astrological principles are applied to a horoscopic chart drawn up for their own birth. The singular accuracy of the data thus generated is often so astounding in its accord with one's own subjective assessment of one's personal characteristics that it is impossible to dismiss it out of hand. This correlation of Astrological interpretations with the skeptic's own acknowledged personal traits is often enough to strongly suggest that such astrological correspondences truly do exist. Further instances of such detailed correspondence in the charts of other well known individuals serve to offer further evidence for the effectiveness of astrological interpretations; and the more one delves into the meanings of the symbolic language of the planets in their ongong transits, progressions, and angular relationships, the more firmly convinced one becomes of the strange and inexplicable correlation that apparently exists between the planets of the solar system and the lives of humans living on the planet Earth.

And yet, we must ask, how could such a correspondence be possible? By what possible means could the positions of the Sun, moon and planets at the moment of an individual's birth constitute the psychological framework of that individual? And how could the angular relationships of the continuing movements of those planets to their positions at birth have the slightest effect on that individual's evolution and development? These are questions that have been asked of the defenders of astrology for the last two millennia; and the lack of any reasonable response, the absense of any verifiable (or even unverifiable) theories to account for the correlations purported by astrologers to exist between planetary patterns and human character, psychology and behavior is the primary reason cited by skeptics for their rejection of the claims of astrology. No electromagnetic-type fields of force have been discovered to account for it; no observable 'planetary rays' seem to be present; no viable theory of universal sympathy or synchronicity has even been put forward. How then account for either a causal or an acausal synchrony between the angular positions of the Sun, moon and planets and the minds and bodies of human beings living on earth?

Astrologers themselves admit they haven't a clue as to how the planets and human consciousness are connected. 'We see the effects of this correspondence,' they say, 'but we do not understand the mode of its efficacy.' And they point out that, of course, the same is true of so many phenomena—like the force of gravity, and the weak, strong and electromagnetic forces. Certainly we know they exist and operate, but no one knows *how* they work or why. We know that gravity exists by observing its effects, though we don't fully understand the mechanism behind it or how to reconcile it with our quantum theories of the microcosm. Likewise, we see the effects of molecular formation in the objects around us and in ourselves; but we don't really understand why the elementary particles form out of the initial explosion of Energy, or where the force comes from that causes them to bind together into atoms, and those into molecules, and from thence into larger living structures. Indeed, what is it that constitutes the life-force of sentient beings? How does it originate? And how does it operate? And what of that most fundamental phenomenon: light? It is a complete mystery. The ambiguous wave-particle duality of light, shown by many different experiments in the scientist's lab, reveals the indefinability of light at the quantum level. We know it is, but we don't know *what* it is or *how* it works, or why its velocity should be a constant 299,792,458 meters per second.

The observed correlation of the positions of the Sun, moon and planets in the solar system with the lives and psyches of the inhabitants of Earth fits right in with all those other unsolved mysteries. We see and experience the connection, though we can't explain the why or how of it. So what else is new? Clearly, the subjective and objective data accumulated tells us that such a correspondence exists. Yet, the critics tell us that astrology *cannot* work in the mechanistic and unensouled universe which science has presented this universe to be. 'In a universe such as contemporary science describes,' they say, 'astrology cannot possibly work!' And, of course, they are right. Therefore, either something is wrong with the astrological idea of correspondences between planets and people, or there is something wrong with the model of the universe which contemporary science portrays. Perhaps the answer to how astrology works must be sought in an entirely different framework of understanding from the usual empirical cause-effect framework in which most of us operate.

Let us look then for some alternative answers from one of our foremost thinkers on the subject of astrology: the author of the watershed classic, *Cosmos And Psyche*, Richard Tarnas. In that book, Tarnas suggests that astrology, from its earliest beginnings, is based on a worldview which he refers

The Divine Universe

to as a "primal" one, a mindset that sees the inner and outer worlds as co-constituents of an all-embracing world-Soul (*anima mundi*) that permeates both cosmos and psyche. In this "primal" worldview,

> The human psyche is embedded with a world psyche in which it complexly participates and by which it is continuously defined. The workings of that *anima mundi*, in all its flux and diversity, are articulated through a language that is mythic and numinous. Because the world is understood as speaking a symbolic language, direct communication of meaning and purpose from world to human can occur. The many particulars of the empirical world are all endowed with symbolic, archetypal significance, and that significance flows between inner and outer, between self and world. In this relatively undifferentiated state of consciousness, human beings perceive themselves as directly—emotionally, mystically, consequentially—participating in and communicating with the interior life of the natural world and cosmos. [1]

This "primal" mindset is contrasted with the "modern" mindset, influenced as it is by the methods and conclusions of the empirical sciences, which assumes a distinct separation between subject and object, between self and world, allowing for no breach of this cognized barrier. The demands of this empirical mindset have taught us to see the world objectively, divorced from human subjectivity, and this perceptive framework has effectively erected a mental defensive barrier against the "primal" worldview.

In the primal (or mystical) worldview, all in the universe is one organic and interrelated whole, and each separate element fits into that whole as an integral component. All things do indeed move together of one accord; not a sparrow falls nor a grain of sand on the ocean's bottom is moved by the currents that is not coordinated with all else in a continuum of Divine interaction. All is contained in the Mind of God, as images contained in a dream are contained and coordinated in the mind of a dreamer. We, who are but insubstantial images in that Mind imagine in our turn that we are, and all about us is, substantial, real, a solid edifice of reality that we can cling to and hang our hats on. But even this body which we label "I" is but a fleeting shadow, a flickering image on a passing screen in the Mind of the One whose bodies all these truly are. It is a dream-world, a projection of a dancing spray of beams upon an infinite expanse of Thought.

In our accustomed "modern" view of a universe of material effects from material causes, all separately isolated from one another, the seemingly dead planets circling the Sun have no bearing whatever on the minds of men on earth. In such a world of independently moving subjects and independently moving objects, how could there possibly be a correlation between the two? Impossible! Inconceivable! But—suppose a world all magically interconnected, made of Thought-streams darting here and there as bits and pieces of reality and force, a dream scenario in which all melds with all—why then, of course the universal Thought-currents directing the movement of the planets and the mental currents in the minds of human beings may blend and press upon each other in an easy way. What happens there is intuited here, sending vibrations through the one expansive gossamer field containing all in this dreamtime play.

In the Indian religious tradition, this phenomenal world is referred to as *Maya*. Maya is the Thought production of the Divine Mind. It is a play of light and energy; this energy (constituting light) forms the substance of Maya. We must grasp it as a Whole, without attempting to reduce it to elementals or causal relationships. Within it there are cohesive forces, but these too are irreducible to separable elements, just as it is impossible to define the elements or cohesive forces within our own dreams or fantasies. There are no individual elements or forces; the universe, like a dream, is a Whole, and operates as a Whole. Under the spell of Maya, we are deluded into believing that we are our bodies and are independent entities separate and distinct from the world of our experience. It is only through an occasional glimpse of clarity that we become awakened to the truth that we live within the one Spirit, one world-Soul, and this body and all nature is His; that it is an illusory world made of projected thought and images in which all things are united and joined in the one all-pervading Spirit. In the delusive world of Maya, where all appears solid and permanent, the planets moving in the heavens are only objects, disconnected from our lives and minds. But, in the revealed world of the living Spirit, there is an active force that interconnects the heavens and the earth and all that exists in one Consciousness, deliberate and entire, guiding and directing every soul and everything in this universe by Its power.

Now, astrologers must possess something of this primal worldview in order to accept and account for the interplay between cosmos and psyche, between world and self, in the universal consciousness that is the *anima mundi*. And it is here, at the most fundamental level, that we discover the great divide between the "primal" and "modern" worldviews, as well as between those who are able to accept and embrace astrological principles and those who are not. From the "primal" perspective, to look for *empirical* proofs for astrological "influences" is irrelevant and beside the point. The connection between planetary positions and human psychology is neither physical nor mental; it is a consonance taking place within the one Spirit—a

The Divine Universe

Spirit or Soul that is both immanent and transcendent, that resides in the individual's innermost being, and yet is all-pervasive, that acts not only as the Providence and guiding Logos of all things and all beings, but as the very Self of those beings. For those of us to whom experience has taught the truth of such notions, the rationale for astrology is thus rendered adequate; and for those to whom such notions are nonsense, astrology must also appear to be nonsense.

Notes:
1.   (Richard Tarnas, *Cosmos And Psyche*, N.Y., Viking Penguin, 2006; p. 17.

24.

THE SOUL OF ASTROLOGY

An astrologer, long familiar with the language of astrological symbolism, can look at the natal horoscopic chart of Isaac Newton and easily discern the primary features of the soul who bore that name, and recognize in these features the historical man; or he may look at the chart of Ralph W. Emerson and discern the soul tendencies impelling that kindly figure, and recognize, as by an interior photograph, the very blueprint of the man's soul. And likewise with every notable character with whom we are familiar: to those conversant with the language, each birthchart is a faultless portrait of the man or woman thus represented. The charts are faultlessly accurate portraits because they represent the cosmic factors involved in the makeup of those souls in the space-time moment of their embodiment. They stand, indeed, as illustrative proof of the interconnection of the soul's qualities and the heavenly environment which accompanies its incarnation.

We are able to see the concentration of genius in the chart of Einstein; we can see the concentration of harmony in the chart of Beethoven; the concentration of madness in the chart of Manson, the concentration of artistry in the chart of Sinatra, the concentration of poetry in the chart of Swinburne, the concentration of Spirit in the chart of the contemporary saint, Amma Mata Amritanandamayi. And we must ask, 'Were all these manifestations of God's life brought to light instantaneously solely by the happenstance of the architecture of the heavens at the moment of their birth?' If we answer "yes" to this question, we have rejected the evolution of the soul, self-effort, karma, and the efficacy of the individual will, and relegated our personhood to the fiat of the stars. We must recognize that these souls had already gone through lengthy preparations and development in previous incarnations, and are now called to make their re-entrance upon the world stage in correspondence with the mirroring spectacle of the heaven's design.

Those souls whose purpose sets them apart, whose aims are strong and focused upon the accomplishment of their destined role, are brought to birth in concert with the starry pattern that portrays their gathered wealth of wit or wisdom or vision or art. All come into life at the intended moment, in concert with the perfect unfolding of the universal array in the Divine Mind and at the behest of the one all-encompassing Soul. "All things move together of one accord; assent is given throughout the universe to every falling grain." The positions and angular relationships of the planets, the necessity of the

The Divine Universe

times, and the appearance of the souls of the great and small, all live and move and exist, entwined together, by that one assent, of that one accord.

One of the greatest seers of the nature of the soul was the Egyptian-born Roman philosopher-mystic, Plotinus (205-270 C.E.). Plotinus had experienced "the vision of God", had ascended in awareness to the transcendent Ground, the absolute Self; and he described in his writings the ascent from body consciousness to God consciousness. He asserts that, in the manifestation of this universe, consciousness moves downward toward limitation, from the One to the Divine Mind to Soul to body (or matter); and it has the power and strong inclination to rise once again from body to Soul to Divine Mind to the eternal One. Consciousness, according to Plotinus, is on a sliding scale from God to matter, from matter to God. We are not separated from God; we live in a continuum (or spectrum) of consciousness, where the pure Consciousness of God rests at a higher, but accessible octave. On that variable scale of Consciousness, we may know ourselves as an individualized soul at one moment, and as the undifferentiated Source at another.

From the vantage point of the One, the Eternal Source, there are no souls; identity is and always was the one Divine Consciousness. But souls *do* exist at a lower 'octave' of Consciousness where physical manifestation occurs, and identity is misinterpreted as associated with the individual body/mind. From this perspective, individualized souls, though manifestations of Consciousness, are illusory. My own experience of the awakening to "higher" levels of Consciousness, and the corresponding absense of the (lower) consciousness of a personal identity or a spatio-temporal presence at that time makes this 'sliding scale' (or graded specturm) theory seem a very plausible explanation.

The ascent of consciousness occurs, says Plotinus, quite unexpectedly in a moment of concentrated awareness focused inwardly. The mind ascends, as it were, to its subtler state, and from there to what Plotinus calls the "All-Soul," all the while drawn on by its inherent thirst to know its Source. When it comes inwardly to a perfect, concentrated stillness, it emerges from its time-bound isolation as an individual creature, and awakes to its identity as a participatory fragment within an all-inclusive creative Power. And yet above that creative Power, at a yet subtler stage of consciousness, it knows itself as the eternal One from which the creative Power takes its origin. It knows This, not as an object is known to a knowing subject, but as the subject's own primary and eternal Identity.

Man, Plotinus asserts, as an evolute of the One, contains within himself all levels of manifestation, from the absolute Unity to the creative Energy, to the soul, to mind, and finally to the gross physical body; and is capable of returning in consciousness to his Origin. It is in relation to man that this out-flowing radiance from subtle to gross is described in the Eastern yogic tradition as well. Man, who is at his center the unqualified Self (*Atman,* or *Brahman*), manifests from the supracausal (*Turiya*), to the causal (*Prajna*), to the subtle or astral (*Taijasa*), and lastly as the gross physical body (*Vishva*).

The levels of human reality, from the gross physical body inward, have been variously named and described; and in all true metaphysical systems the primary teaching has been that one is able to reach to and experience that Self by way of the inner journey only, seeking it by way of self-examination, purification, contemplation and selfless devotion. Self-examination reveals to us that we are more than the physical body with which the immature consciousness identifies. We are more than the effusive mind with which some others identify, more than the intellect which reasons and oversees the mind, more than the individual soul which, through purification, evolves from lifetime to lifetime.

The soul, seeking God, scans the inner darkness, as though to discover another, as though awaiting something external to itself to make its presence known. But as the concentration focuses within, the mind becomes stilled, and suddenly the seeking soul awakes. No external has made its appearance; it is the soul itself, no longer soul, which knows itself to be the All, the One. Like a wave seeking the ocean, the seeker discovers that it is, itself, what it sought. Through contemplation and selfless devotion to that highest Self, we discover that we are the Life in all life, the integrated Whole of which all manifest creatures and things are a part. And, at last we awake to the supremely ultimate Identity, knowing ourselves as the one Light of existence, the Source of all manifestation, the one God who is the true Self of all, and from Whom all else follows.

From the standpoint of the human experience, these various levels of being are not clearly separated off from one another with clear demarcations to indicate where one ends and another begins, but tend to merge one into the other in a gradual and vaguely perceived manner. We are aware of being identified with one or another level of Being according to the activities which follow upon it. When we are identified with the physical body, we are operating almost solely through our senses, and we find our gratification in things of sense. When we identify with the mental realm, we are conscious of the inner play of random thoughts and images, and we delight in the play of thought. When we ascend a bit to the intellectual realm, we identify with the critical intelligence which discriminates, censures, and deliberates; thereby elevated in concentration above the rambling mind, we take pleasure in the clarity of discernment. Above this intellect, we experience our soul, at its lower level the repository of our karma, and at its higher level the bearer

The Divine Universe

not only of our highest moral directive and purpose, but the driving impetus guiding us toward our own Source with a heartfelt longing, like that of a moth to a flame. The soul is drawn to the Light within it, [1] and looks, not below to the realm of mental activity or the realm of sense, but above toward the Divine whence it comes.

Those who have risen yet higher (or more inwardly) toward their Source have experienced themselves no longer as individual separate identities, but rather as ideational wave-forms on the one integral ocean of Cosmic Energy. They no longer identify with the composite of body, mind, and soul, but know themselves as having their real identity in the entire undivided ocean of creative Energy in and on which these temporary forms manifest. The conscious awareness focused on this clear vision of the subtler level of its own reality then moves forward, as one moving through a fog comes to a clearing where the fog is no more, to the ultimate and final level of subtlety, the Divine Source, the Unmanifest. Then, it knows the pure unqualified Consciousness that is the Father, prior even to the creative Power which acts as creator; and it knows, "I and the Father are one."

From that vantage point in Eternity one sees one's own creative Power manifesting all that has manifest existence in a cycle of creation and dissolution. There is a bursting forth, just as the spreading rays of the Sun burst out from their source, and then a returning to that source in a cyclic repetition, much as the cycle of the breath's inhalation and exhalation. One witnesses this from that transcendent vantage point, aware of one's Self as the Eternal One, totally unaffected and unaltered by the expansion and contraction of the out-flowing creative Force—as a man might watch the play of the breath or the imagination without being at all affected by its rise and fall. That One is the final irreducible Reality, and It is experienced as identity. Nothing could be more certain than the fact that It is who one really is, always was, and always will be.

Soul, for Plotinus, is an outpouring of the Divine Mind, a living radiance which fills the cosmos and manifests as individual souls. Here is how he describes the Soul:

> There is one identical Soul, every separate manifestation being that Soul complete.[2] ...[The separate manifestations] strike out here and there, but are held together at the source, much as light is a divided thing upon earth, shining in this house and that, while yet remaining uninterruptedly one identical substance. [3]
> ...This one-All, therefore, is a sympathetic total and stands as one living being ...[4] [It] is a Soul which is at once above and below,

attached to the Supreme and yet reaching down to this sphere, like a radius from a center.[5]

...The Soul's nature and power will be brought out more clearly, more brilliantly, if we consider how it envelops the heavenly system and guides all to its purposes: for it has bestowed itself upon all that huge expanse so that every interval, small and great alike, all has been ensouled. ...Each separate life lives by the Soul entire, omnipresent in the likeness of the engendering Father, entire in unity and entire in diffused variety. By the power of the Soul the manifold and diverse heavenly system is a unit; through Soul this universe is a god. And the sun is a god because it is ensouled; so too the stars: and whatsoever we ourselves may be, it is all in virtue of Soul...[6]

What does it mean to say that the universe is "ensouled"? It means that all exists within the Mind of God, and partakes of the essence of God. That presence may be thought of as a universal Soul, or *anima mundi*, which enfolds, inheres in and embodies every element of this cosmic appearance. It is a unified Spirit in which all exists, and by which all constituent appearances are permeated and governed—just as, in our own experience, are all images contained, permeated and governed by the mind in which they appear. It is in such a conception of the universe and the Soul that we find the possibility of a correspondence between existing planetary patterns and the incarnation of individual souls, the sum of whose karmic histories are depicted in those patterns. Indeed, such a miraculous correspondence *requires* a universe that is ensouled, a universe in which all things move together of one accord, in which assent is given throughout to even the most insignificant occurrence. I believe that, the more we examine our own lives and the nature of our cosmos, the more we shall come to perceive our own Divinity and the Divinity of our own miraculous universe, where souls manifest, evolve, contribute, and come eventually to know their own identity with the one Self of the universe.

Notes:

1. "From within or from behind, a light shines through us upon things, and makes us aware that we are nothing, but the light is all." (from Ralph Waldo Emerson, "The Over-Soul", *The Works of Ralph Waldo Emerson,* Tudor Publishing Co., p. 174).

2. Plotinus, *Enneads,* IV.3.2: *Problems of The Soul (1).*

3. *Ibid.*, IV.3.3-4: *Problems of The Soul (1).*

4. *Ibid.*, IV.4.32: *Problems of The Soul (2).*

5. *Ibid.*, IV.1.1: *On The Essence of The Soul (1).*

6. *Ibid.*, V.1.2-3: *The Three Initial Hypostases.*

25.

ASTROLOGY AND FREE WILL

All this world is one gigantic motion, one supreme life flowing endlessly through an infinite diversity of lives. If all this be true, is it a superstition to believe that this cosmos is one creation and that each of its parts is bound to all the other parts by mysterious electrical and magnetic bonds of sympathy? Accept this and it is only reasonable to become an astrologer, for astrology is that science of the ancients which teaches of the effect of cosmos as environment upon the lives of all those creatures who exist within this environment.

— *Manly P. Hall* [1]

Part One

Some of the keenest intellects of many early civilizations recognized the correlation between the changing positions and patterns of the planets in our solar system and the changing mental and physical conditions of life on earth. As they learned by observation of the distinct nature of the effects associated with each planet, they ascribed to each a specific kind of influence, considering each of the heavenly bodies, including the sun and moon, to be embodiments of divine powers, or "gods". These gods were both benevolent and mischievous, bestowing both blessings and calamities upon earth and her inhabitants. Each had its own personality and characteristics, and dealt with men on earth in ways compatible with their separate natures.

Today, of course, these beliefs are regarded by many as mere primitive superstitions, having no basis in fact whatsoever. But let us not be so hasty in our judgment of these early mythologizers. Over the centuries, the correlations between planetary patterns and specific psychological and physical effects on Earth have continued to be chronicled by observers of the heavenly dynamics, and much evidence has been accumulated to show a factual basis for these planetary myths of correspondence. Today, the notion of "gods" is frowned upon; instead, we like to call those various distinct energies associated with the planets "archetypes", after Carl Jung, who utilized the term (originally coined by Plato) to speak of those intangible influences. This may also prove in time to be an inadequate term; but for now, we shall speak of the power of the gods as "archetypes".

The Divine Universe

For a long time now, this study of the correspondence of the archetypal energies associated with the planets and the patterns of mental and physical changes on earth has gone by the name of *astrology*. It is fashionable among those "learned" in the universities to regard this study as having no scientific basis and as being merely a throwback to superstition and ignorance, appealing only to the indiscriminant and gullible masses. But we should remember the words of the great astrologer, Ptolemy, who warned, "It is a common practice with the vulgar to slander everything which is difficult of attainment." [2] How, then, shall we define these *archetypes*? Here is what philosopher, astrologer, and author of the highly regarded affirmation of astrological principles, *Cosmos and Psyche*, Richard Tarnas, says:

> Archetypes can be understood and described in many ways, and in fact much of the history of Western thought from Plato and Aristotle onward has been concerned with this very question. But for our present purposes, we can define an archetype as a universal principle or force that affects—impels, structures, permeates—the human psyche and human behavior on many levels ... Moreover, archetypes seem to work from both within and without, for they can express themselves as impulses and images from the interior psyche, yet also as events and situations in the external world.
>
> Jung thought of archetypes as the basic constituents of the human psyche, shared cross-culturally by all human beings, and he regarded them as universal expressions of a collective unconscious. Much earlier, the Platonic tradition considered archetypes to be not only psychological but also cosmic and objective, as primordial forms of a Universal Mind that transcended the human psyche. Astrology would appear to support the Platonic view as well as the Jungian, since it gives evidence that Jungian archetypes are not only visible in human psychology, in human experience and behavior, but are also linked to the macrocosm itself—to the planets and their movements in the heavens. Astrology thus supports the ancient idea of an *anima mundi*, or world soul, in which the human psyche participates. From this perspective, what Jung called the collective unconscious can be viewed as being ultimately embedded within the cosmos itself.
>
> ... The basic principle of astrology is that the planets have a fundamental cosmically based connection to specific archetypal forces or principles which influence human existence, and that the

patterns formed by the planets in the heavens bear a meaningful correspondence to the patterns of human affairs on the Earth. In terms of individuals, the positions of the planets at the time and place of a person's birth are regarded as corresponding to the basic archetypal patterns of that person's life and character. [3]

According to this interpretation, the natal chart represents the psychological make-up or orientation of the individual at birth; and the ongoing progressions and transits reflect the changing modes of thought and experience occurring throughout ones present life. The position of the transiting planets therefore represents a sort of evolving map of the intricately changing network of our mental experience. But there must inevitably arise the question of *how* and by what means are the changing positions of the planets synchronously related to the human psyche. The changing aspects of transiting planets to the positions of planets existing at birth can clearly have no effect on a person unless those natal planetary positions are an integral part of the makeup of an individual's personal psyche. It seems that the position of the planets at birth is somehow imprinted on that soul, or is in one way or another synonymous with the characteristics of that particular individual's psyche; so that, the transiting aspects to the planetary positions of the natal chart are relating to something integral to the individual. They are relating to the living psyche of the individual, which in turn is synonymous with the planetary arrangement existing at his/her birth.

If this interpretation of "planetary correspondences" is correct, then every individual born bears within itself the imprint and structure of the planetary arrangement existing at that very moment, and is in fact an embodiment of that planetary arrangement. And the movements of the planets, along with their changing relations to one another, during the course of the life of the individual are intimately correlated with the unfolding life and psyche of that individual. It is not that one is considered to be causing the other; they are regarded instead as merely correlated events in the universal unfoldment. They are merely two synchronous mirror images of the activity of the *Nous*, the Divine Mind. Here, again, Richard Tarnas, on why astrology works:

> It seems unlikely to me that the planets send out some kind of physical emanations that causally influence events in human life in a mechanistic way. The range of coincidences between planetary positions and human existence is just too vast, too experientially complex, too aesthetically subtle and endlessly creative to be explained by physical factors alone. I believe that a more plausible

and comprehensive explanation is that the universe is informed and pervaded by a fundamental holistic patterning which extends through every level, so that a constant synchronicity or meaningful correlation exists between astronomical events and human events. This is represented in the basic esoteric axiom, "as above, so below," which reflects a universe all of whose parts are integrated into an intelligible whole. [4]

In dealing with astrological "influences" one needs, therefore, to take a universal all-inclusive perspective, and to recognize the truth of the fundamental dictum that "all things move together of one accord." From this perspective, the universe is the manifestation of the one Intelligence, the *Nous* or *Logos*; all is one integral life in which every entity and every action is interrelated, functioning as coordinated aspects of the universal expression. In such a view, the planets are merely "signs", indicators of prevailing influences or energies currently operating, and have no causal function. This view, also, asserts a marvelously complex and exquisite perfection in the unfolding of the universe, and underscores the existence of a Divine Intelligence in operation down through each member, upholding, activating, and supporting all. The individualized soul, the result of its previously created karma, comes into the world at exactly the moment that the planetary arrangement mirrors the qualities of its being. What a truly extraordinary wonder of Divine creative perfection!

But should we gather, then, that we are wholly governed by these planetary energies (archetypes)? That there is a cosmic determinism at work here that is inescapable? That our sense of individual freedom is merely an illusion, and that we must unwittingly and unerringly follow the cosmic fiat as inscribed in the movements of the stars? And, perhaps most importantly, if there is, instead, a means by which each individual soul possesses a free and indetermined will, quite beyond the "meaningful correspondence" that exists between astronomical and human events, what is the explanation for such a free will?

We must believe in free will; we have no choice!
—Isaac B. Singer [5]

Part Two

The soul, or psyche, of each individual, though embodying the cosmic arrangement at the moment of its birth, and constituting the specific tenor and structure of the life of the individual, has at its core the eternal Consciousness which is the principle and primary element of its being, constituting its permanent Ground and Self, beyond all projected energies resulting from any temporary arrangements of the cosmic array. Therefore, the cosmic arrangement at the moment of our birth into this universe may constitute our temporal identity; but the One who projects this universe, and in whose Mind we exist, constitutes our eternal Identity. This eternal Identity remains throughout our existence, and is unaffected by any transient conditions, such as the planetary patterns of relationship appearing within the manifest universe.

The Neoplatonic conception, as put forward by Plotinus (205-270 C.E.), as well as the Vedantic conception, put forth in the Upanishads, satisfactorily explains this eternal Principle of freedom. The Divine Mind (*Nous* or *Brahma*), which is the active element of the Divine Consciousness, projects a coordinated Dream-world of immense vastness and complexity (the manifested *Cosmos* or *Maya*). Yet the source and heart of all existence, the substratum of Divine Consciousness, the Ground of the Soul (*the One* or *Brahman*), remains constant. It is independent of and unaffected by this surface play of universal phenomena; for the world of physical and mental phenomena is but an appearance, a sort of superimposition, on this substratum of Divine Consciousness. For most of us, the mind's continuous display of this superimposition of both physical and psychological states synchronous with the positions and angular relationships of the planets is extremely persuasive, becoming the primary basis of our psychologically perceived reality. But, through deep meditation or deliberate recollection, we are able to maintain identity with the Conscious substratum of Being, and able to view the ongoing parade of transient physical and mental conditions and images related to existing planetary energies as but the superimposed activities of that Conscious substratum.

Therefore, when we consider the correlation of planetary events and human events, we are not dealing with a straightforward cause-effect relationship. This is because we humans are of a two-fold nature; we are, in essence, identical with the divine Consciousness, our Divine Self, which assures us of a free will; and we are only secondarily products of the creative Power (*Nous* or *Brahma*) which begets the material body-mind complex along with its accompanying karmic tendencies. The winds of all the influences of all the planets may blow, but the Divine Self may yet remain unmoved, withholding and denying her consent to the influential powers; or better, she may use the influences of those planetary powers to her own Divine purposes, rather than to the merely pleasurable mental, physical and emotional activities to which they tend to incline. Conversely, if an individual's sense of the Divine Self is weak, the individual's will may be swayed by the mental and physical influences impinging on her, and surrender to their power. But, with a determined dependence on and identification with the Divine Self, the individual will has the free and final word on the course of the life it rules.

We are a combination, a duality, of identities: we are the Divine Self (*the One, Brahman*), and we are also the projected (superimposed) individual soul (*jiva*). Our essence, the one Divine Consciousness, is the only true 'I' in all the universe and beyond; It is everyone's eternal Identity. But, by His mysterious Power of illusion (Maya), each body born in this world takes on a limited set of characteristics as well, which constitutes one's limited temporal identity, otherwise known as the *jiva*, or individualized soul. According to that soul's previous mental tendencies, and in synchrony with the evolving motions of the planets and celestial bodies as they relate to the place on Earth where that soul takes birth, the characteristics of each soul are made manifest. The astrological interpretations of the planetary positions at one's birth can therefore be helpful indications of the soul characteristics of each person born.

The astrological natal chart is an authentic diagram of the individualized soul, but it says nothing of the Divine Identity, or Self, underlying the manifestation of that soul. The 'soul' is in essence the Divine as it appears within the dream-fabric of *Cosmos/Maya*. It partakes of both the Divine and the illusory—just as in a dream, we partake of both our true conscious selves and an illusory self. The analogy is exceedingly apt, as in both instances, we retain our fundamental reality, while operating in an illusory 'imaged' reality. The individual soul (or *astral* body), as portrayed in the astrological chart, is, to a great degree, who we are; and we operate in this life from the past karmic tendencies embodied in that natal chart. However, at a more fundamental level, we are identical with the Divine Self, which comprises our freedom to will and act from a level of consciousness beyond our soul properties and characteristics. The past karmic tendencies are very powerful in their influence; and they can lead us astray, unless we are able to identify with the Divine Self and turn those inherent tendencies to Divine purposes.

A recent example will suffice to illustrate this dual identity: A young man, a college student, named Seung-Hui Cho, went on a recent rampage, killing thirty-two of his classmates at a Virginia College. The young man's natal chart shows the difficult karmic limitation suggested by the Sun's square aspect to a close conjunction between Mars and Pluto. A predictably volatile and violent aspect indeed! Also, at the time of his birth, Jupiter was in exact conjunction with Neptune. Such natives have a tendency, if there are other challenging factors, to "lose contact with reality and live in a world of private fantasy". This natal chart describes the soul characteristics under which this young man took birth. They were not conditions which were imposed from without; they were conditions previously forged in his own soul, and they describe the embedded tendencies (as depicted in the natal chart) which constituted the framework of his recent life.

But underneath this projected framework there stood the divine Consciousness, the free Will of the Self. Would he identify with that higher Soul Essence and be triumphant in overruling the limiting structure of his accumulated tendencies, or would the tendencies win out? We now know the terrible answer to that question. But we must acknowledge that, despite the overwhelming strength of the negative tendencies embodied in this soul, at his core, he was free to refuse assent to their promptings. The negative soul tendencies won out. They proved too deeply entrenched, too overwhelmingly reinforced in this present life, to be overcome; but we must never doubt that, in his essential Being, he was free to choose. "The fault, dear Brutus, is not in our stars, but in ourselves." [6]

Here is that venerable sage, Plotinus, with some pertinent comments on this subject:

> If man were... nothing more than a made thing, acting and acted upon according to a fixed Nature, he could be no more subject to reproach and punishment than the mere animals. But as the scheme holds, man is singled out for condemnation when he does evil; and this with justice. For he is no mere thing made to rigid plan; his nature contains a Principle apart and free. [7] ... This, no mean Principle, is... a first-hand Cause, bodiless and therefore supreme over itself, free, beyond the reach of Cosmic Cause.[8] ...In [Plato's] *Timaeus* the creating God bestows the essential of the soul, but it is the divinities moving in the Cosmos [i.e., the planets] that infuse the powerful affections holding from Necessity—our impulse and our desire, our sense of pleasure and of pain—and that lower phase of the soul in which such experiences originate. By this statement our personality is bound up with the stars, whence our soul takes

shape. And we are set under necessity at our very entrance into the world: our temperament will be of the stars' ordering; and so, also, the actions which derive from temperament, and all the experiences of a nature shaped to impressions.

...[But] there is another [higher] life, emancipated, whose quality is progression towards the higher realm, towards the Good and Divine, towards that Principle which no one possesses except by deliberate usage. One may appropriate [this Higher Principle], becoming, each personally, the higher, the beautiful, the Godlike; and living, remote, in and by It—unless one choose to go bereaved of that higher Self and therefore, to live fate-bound, no longer profiting, merely, by the significance of the sidereal system but becoming as it were a part sunken in it and dragged along with the whole thus adopted. For every human Being is of a twofold character: there is that compromise-total [consisting of soul conjoined to body], and there is the authentic Man [the divine Self]. [9]

And here is a selection from the *Svetasvatara Upanishad*, which expresses, in it own way, much the same realization:

I sing of Brahman: the subject, the object, the Lord of all!
He's the immutable Foundation of all that exists; those souls who realize Him as their very own Self are freed forever from the need for rebirth. When that Lord, who pervades all the worlds everywhere, gave birth to the first motion, He manifested Himself as creation. It's He alone who is born in this world. He lives as all beings; it's only Him everywhere.
... Those who have known Him say that, while He manifests all worlds by His Power, He remains ever One and unchanged. He lives as the one Self of everyone; He's the Creator and Protector to whom all beings return. The Lord is the Foundation of both aspects of reality: He is both the Imperishable and the perishable, the Cause and the effect. He takes the form of the limited soul, appearing to be bound; but, in fact, He is forever free.

Brahman appears as Creator, and also as the limited soul;
He is the Power that creates the appearance of the world.
Yet He remains unlimited and unaffected by these appearances.
When one knows that Brahman, then that soul becomes free. The forms of the world change, like clouds in the sky; but Brahman, the

Lord, remains One and unchanged. He is the Ruler of all worlds and all souls. Through meditation on Him, and communion with Him, He becomes known as the Divine Self, and one therefore becomes freed from illusion. [10]

It is important to have a clear understanding not only of one's Divine Ground, and to identify with one's eternal Freedom, but one should also have a complete understanding of the makeup of one's soul as indicated by the planetary pattern existing at birth, as well as of the nature and occurrance of the various changing planetary conditions as they manifest daily in our lives. An awareness of the archetypal energies currently prevailing in one's life gives an extraordinary advantage in the timing and utilization of those specific energies for the enfolding of one's potential to manifest and express the freedom of the Divine Will. As Richard Tarnas explains,

> Astrology can serve to greatly increase personal freedom ... Partly this is because awareness of the basic archetypal structures and patterns of meaning in one's birth chart allows one to bring considerably more consciousness to the task of fulfilling one's deepest potential, one's authentic nature. But [also because] the more deeply we understand the archetypal forces that affect our lives, the more free we can be in dealing with them. If we are altogether unconscious of these potent forces, we are like puppets of the archetypes; we then act according to unconscious motivations without any possibility of our being intelligent agents interacting with these forces. To the exact extent that we are conscious of the archetypes, we can respond with greater autonomy and Self awareness. [11]

The great Vedantic sage, Shankaracharya, taught, "the soul is none other than Brahman" (*jiva brahmaiva naparah*). And this is true; in essence, the soul is identical with the transcendent Source of all, and is supremely, absolutely, free. In its transcendent aspect, it is always free, immutable and unaffected by the bodily conditions or worldly circumstances of individuals; however, when the soul identifies with the conditional, it is bound; it is subject to being carried along in the floodwaters of the archetypal forces of Nature. Only when it knows and identifies with the One, the Divine Self, does it realize and manifest its true freedom. This is the view of Vedanta, and the basis for its concept of "liberation"; and this is the view of Plotinus as well.

Another great seer and teacher put it well when he said, "You shall know the Truth, and the Truth shall make you free." According to

The Divine Universe

this understanding, a man is free insofar as he is cognizant of his essential identity with the Highest, and bound when he departs from the knowledge and awareness of his Divinity, identifying with the body/mind complex. He then succumbs to the rule of earthly necessity, and is moved willy-nilly by the causative forces inherent in Nature. He has the power, as the Divine Self, to will freely, unencumbered, uncompelled by circumstance; and, for that reason is responsible for his individual actions. All souls are linked by inclusion to the one Soul, and by extension to the Divine Mind; but only he who is cognizant, aware, of his Divine Identity, is truly free.

Notes:

1. Manley P. Hall, *The Philosophy of Astrology*, Philosophical Research Society, Inc., 1947; p. 35.

2. *Tetrabiblos*, Book I, Ch. 1, from J.M. Ashmond, *Ptolemy's Tetrabiblos*, Chicago, Aries Press, 1936; p. 1.

3. Richard Tarnas, Ph.D., *An Introduction to Archetypal Astrolological Analysis*, pp. 2-3.

4. Richard Tarnas, Ph.D., *Ibid.*, pp. 3-4

5. A saying attributed to Isaac B. Singer (1904-1991), the Nobel prize-winning author of Yiddish stories and novels.

6. William Shakespeare, *Julius Ceasar*

7. Plotinus, *Enneads, III. 3.4: On Providence [2]*

8. *Ibid., III. 1.8: Fate*

9. *Ibid., II. 3.9: Are The Stars Causes?*

10. *Svetasvatara Upanishad, I.7-10, II.15*

11. Richard Tarnas, Ph.D, *An Introduction to Archetypal Astrological Analysis*, pp. 2-3

26.
ETERNAL FREEDOM

Now that we have firmly established that you possess in your very nature the capability of willing freely, let's take a look at that free will from a more expanded perspective: This world, and all that is in it—indeed this entire universe—exists in the Mind of God. This far-flung cosmos is a Mind-born image constituted of Thought. It is made of the Consciousness of the one Divine Self. It is that very Consciousness that we partake of as we become aware of our own existence. That Consciousness, manifest as us, is the inner sense of 'I am' that constitutes our awareness, our identity. The immense drama taking place as this universe, therefore, is, in many respects, like a dream. In this analogy, God is the Dreamer, we are the characters in the dream. Yes, indeed, we are able to will freely in this dream, as we are essentially identical with the Dreamer, partaking of His utter freedom.

To understand this better, let's look at our own dreams: In our dreams, our dream characters exist as images in our own minds, borrowing their awareness from the consciousness in which they live and move and have their being. Though they experience a freedom of movement and choice, it is the freedom of the dreaming mind (ourselves) that is the foundation of that sense of freedom. Theirs is but an imagined freedom; they are in fact entirely governed by our own subconscious willing. When we awake, the dream characters vanish, and we alone remain. Similarly, we, in this phenomenal reality, experience a freedom of movement and choice, but it is the freedom of the One in whom we exist that is the foundation of that sense of freedom. When God withdraws this dreamlike universe of phenomena back into His own Consciousness, we vanish, and He alone remains.

In our dreams, it is always only our 'real' selves who truly exist; the dreams are but images playing in our own minds. Likewise, in this phenomenal universe, it is always only God who truly exists; the universal phenomena are but images playing in His own Mind. It is He who is the only Existent when the universe is imaged forth, and He is the only Existent when the universal image is extinguished. He is always the only one who exists. And so, the free will that we experience during our brief imaginary existence is really *His* freedom of will. There is really no separate 'I' to claim possession of such a will, and there never was. The only 'I' was His all along. And all along, the freedom that we claimed was, and shall always be, His. The bottom line is, yes, we have free will! But we exist in Him, and what we think is ours is truly His. 'I' and 'Him' are ultimately not two; and so there is no contradiction here. Be free, and know that you are Him!

༄ ༄ ༄

27.

APPENDIX 1: REFLECTIONS ON THE TWO DEFINITIONS OF ENERGY

After thinking for some time about the two definitions of the word, *energy*, as embodied in the definitions of David Darling: ("1. A measure of the ability to do work—for example, to lift a body against gravity or drag it against friction, or to accelerate an object. 2. An intrinsic property of everything in the universe, including radiation, matter, and, strangely enough, even empty space."), I began to see that these two definitions are intimately related. Before I get into that, however, let me articulate the apparent differences between these two conceptions of energy.

Richard Feynman, who was a highly respected and thoughtful physicist, said in one of his lectures, "It is important to realize that in physics today, we have no knowledge of what energy is." Many will find this idea preposterous; but let me attempt to explain the rationale behind Mr. Feynman's statement that "we have no knowledge of what energy is": First, let me explain that there are two different but equally valid definitions for the word *energy* currently in use: The two definitions are related, one applied by physicists, in a context limited to isolated systems; the other, of popular origin, applicable in a general, universal framework. In the 1st definition, "energy is the capability of a system to produce work". But this definition gives *energy* no specific measurement value; it takes on a specific value only when associated with a particular "system". In other words, energy, by this definition, has no precise existence as a stand-alone entity, but must be coupled with a particular system in order to have meaning. For example, *electrical energy* is measured in electron volts (eV); motion (*kinetic energy*) is measured in joules; *heat energy* is measured in calories; and so on. Each of these values are quite different, and apply only to each specific system. So, perhaps what Feynman realized was that the term, *energy*, by itself is meaningless—other than as a category: the act of producing an effect! Only when applied to a particular system, and coupled with a particular unit of measure, does it take on a specific meaning. Here is the way this somewhat complicated idea is expressed in two college textbooks on Thermodynamics: "*Energy is a mathematical abstraction that has no existence apart from its functional relationship to other variables or coordinates that do have a physical*

The Divine Universe

interpretation and which can *be measured."* [1] ... *"Although no simple definition can be given to the general term energy, E, except that it is the capacity to produce an effect, the various forms in which it appears* can *be defined with precision."*[2] Once this is understood, it is easy to see why Feynman intimated that *energy* (in the 1st definition) is still an indefinite term.

The 2nd definition of energy has been up to now only implicit in popular literature: here it is a noun denoting the underlying elemental sea of activity from which all matter manifests and of which all consists. This Energy is not an amorphous, featureless entity, but contains and includes various specific 'kinds' or manifestations of energy such as "electrons", "photons", and "quarks". These specific kinds of energy, being themselves irreducible, are nevertheless manifestations of Energy, being nothing less than or other than Energy. Though they are formless, they have separability; in other words, they are individually discreet charged "fields" of Energy. It is only when the quarks get together to form protons and neutrons, and these join with electrons, forming atoms; and these conglomerate bundles of Energy (atoms) join to form molecules, that we have the appearance and tactibility of substance–of matter. But, in fact, there is only Energy—Divine Energy. This 2nd definition of energy therefore is, as stated in the Essay, "What Is Energy?": *"Energy is the elemental creative force, responsible for the manifestation and proliferation of universal phenomena, including matter, motion, force, heat and radiation."* Indeed, it constitutes everything in our universe. Here, again, from the perspective of empirical science, Feynman's words are accurate.

Those with a background in science will no doubt be perplexed and confounded by this 2nd conception and definition of Energy. It is likely that they have been so thoroughly conditioned by the science courses they attended in the schools that they find it difficult even to entertain a revision of the concepts and definitions therein committed to memory. "Energy" has been used (in accord with the 1st definition) since the nineteenth century to describe the capability of muscle, iron, chemicals, steam, etc. to perform actions; and these actions were measurable in terms of joules, calories, coloumbs, etc. Yet, by the mid-twentieth century, as our knowledge of the constituency of the various substances which we call 'matter' became known, and also as a result of our becoming culturally accustomed by Einstein's formula to view mass and energy as mutually convertible, it came to be understood that "energy" is the very essence of everything that exists. It is not merely the measurable actions performed by moving bodies that constitutes "energy"; the bodies themselves consist of energy! Matter itself came to be understood as a mere appearance of solidity created by the subatomic interactions between variously charged wave/particles (quanta) which themselves were reducible to pure Energy.

Appendix

When we go back to the beginning of the universe, prior to the formation of particulate matter following the Big Bang, all of the material universe that now exists in so many diverse forms was then simply Energy (though physicists prefer the relativistic term "mass-energy"). What we see today as 'the universe' is still nothing but Energy; it's just that it has continuously organized itself into ascending levels (or clusters) of apparent substance and solidity over these 15 billion years—but at bottom it is still only Energy. Nonetheless, we must acknowledge that it is a very extraordinary thing—this Energy—that it is able to become an entire multiformed and multiactive universe. And it is able to do this because it is the Energy of Thought originating in an all-powerful and universe-transcending Mind or Divine Consciousness. What we know as "Energy" is really the manifestation of Divine Thought. Think of the origin of this universe: From an inexhaustible yet imperceptible Fountainhead an infinite flow of Energy suddenly manifested, the space-time continuum forcefully spreading along with the expansion of this Energy, as it cooled to become conglomerate bodies of 'matter' spread throughout a universe of light. Can there be any doubt as to the Source of this Energy?

There are, of course, several different versions of the Creation story from a religious perspective. The ancient Hebrew conception of Creation, as expressed in the Biblical book of Genesis, involves a Creator who is other than His creation, who exists wholly outside His creation, in a relationship to His creation similar to that of a watchmaker to a watch. However, in the ancient Indian conception, explicitly formulated in the *Upanishads* and the *Bhagavad Gita*, while God is still other than His creation, His creative Intelligence is inherent in and intrinsic to the initiation and evolution of His creation. The initial projection of His conscious Thought-Energy contains in itself the evolving dream-universe in its entire manifestation, from beginning to end, from the plasmic Energy expanding from the 'Big Bang' to universal form and structure, to the appearance of life-forms, to the development of human receptacles capable of attuning with and intimately experiencing the Divine Consciousness that is their essence.

In the Judeo-Christian Bible, God is depicted as creating His universe in stages—the light; the firmament; the earth; the Sun, moon and stars; the fish and birds; reptiles, land animals and humans, prior to resting from these labors. One may well dispute whether or not this accurately recapitulates the order of appearance of these features of the various stages of creation; but I believe, based on my own visionary experience, that the creative act was a single act synonymous with the sudden singular appearance of the universal Energy at the moment known as' the Big Bang'. I believe that, because the Divine Consciousness imbues His creative Energy with His own life and intelligence, life and intelligence are emergent factors in the unfolding of the manifest universe, requiring no separate and distinct act on the part of God to initiate life on earth. I believe the subtle levels of human consciousness—the astral, causal, and super-causal—exist inherently also in the cosmic dream, linking human consciousness back to the one Divine Consciousness that is God. All is His Mind-production, and all leads back to Him, as a multifaceted dream must find its single source in the dreamer.

I am doubtful that science will be able to trace the acts of conscious evolution in the fossils and artifacts of time, or measure the depths of God's ever-ebullient joy in the hearts and minds of men and women throughout the ages. But perhaps the sparse hints contained herein regarding His purpose and His methods will bring to both men of faith and men of science some modicum of clarity and delight.

Now, the questions arise, 'How can we have two quite different definitions of the one word, *energy*?' 'How can they both be true?' 'And how can these two opposing perspectives on energy be reconciled?' The answer is that the *energy* of the 1st definition, that appears in so many varied forms and modes of activity, is a limited expression of the one ubiquitous *Energy* that is omnipresent, that is the universal Creative Force described in the 2nd definition. We have a 'special' application of the word, *energy*; and we have a 'general' application of the word, *Energy*, conveniently differentiated by the capital first letter. In both cases, the word refers to the same Divine essence; it is just that, in the 1st instance, it refers to a limited and measurable manifestation, and in the 2nd instance, it refers to its unlimited and immeasurable manifestation. The 1st definition of energy (the special definition) is useful technologically; it allows for the measurement of work, force and power. The 2nd definition (the general one) is all-inclusive and therefore of no apparent practical use technologically—which is no doubt why empirical science has ignored it. However, the comprehension and understanding of the underlying fundamental reality of which everything is constituted certainly has a valuable and satisfying psychological, philosophical, and spiritual use. It provides us with a perspective by which all that we see and experience in this life is filled with Divinity, as it truly is. So, we need no longer be confused about the definition and use of the word, *energy*, nor about the definition and use of the word, *Energy*. They are but two references, from two different perspectives, to the one amazing creative Power of God.

Energy is the primary manifestation of Divinity, from which all this universe—both noumenal and phenomenal—is formed. All limited forms of energy partake of the Energy that initiated this universe. All instances of the production of work or heat utilize this original Energy, the sum of which remains constant despite any transformations which might occur.

Appendix

Without that original appearance of Energy (the Big Bang), the Laws of Thermodynamics would have no meaning. What would be conserved? What would provide heat or work? The very existence of matter and life depend upon that initial Energy. Energy is not "the capacity for work"; the capacity for work is simply one of the many qualities or characteristics of Energy. For the frog in the pond, energy is "the ability to jump"; for the physicist, energy is "the capacity for work"; but for the wise who see the larger picture, these are but specific instances of the many manifestations of Energy. If we must define Energy, we must acknowledge that it is the substance of all that is. It is the very breath of God that appears as this far-flung cosmos. Because of that Energy you are able to lift your finger, to blink your eye, to draw your breath. For Energy is the unimaginable immensity of light that shone from God, and metamorphosed into the form and substance, along with extension and duration, that constitutes the entire world of our experience.

I think that eventually all people must gradually come to see and accept the universe as a manifestation of Divine Energy, of Divine Intelligence. For it is the truth of the nature of our world and our own being. Yet, I wonder if anyone reading this book in the 21st century can possibly imagine what a transformation is to be wrought in the vision, in the sciences and the philosophies, in the thoughts and actions, and in the lives of all the people of the world as such a transformed world-vision becomes the norm. Surely all our blindness will be cleared away and, freed of all darkness, our eyes will open to the dawning effulgence of our own eternal Self!

Notes:

1. M.M. Abbott and H.C. Van Ness, *Theory and Problems of Thermodynamics*; Schaum's Outline Series in Engineering, McGraw-Hill Book Company.

2. V.M. Faires and C.M. Simmang, *Thermodynamics,* McMillan Publishing Co., Inc.

3. It must be noted that science does not recognize such a thing as "pure Energy", just as it does not recognize such a thing as "pure matter". For there is no fundamental duality between Energy and matter. If "matter" is simply (Divine) Energy manifest as form and substance, how do we distinguish one from the other? Where would one draw the line between them? Is a photon (a massless quanta of light) matter or Energy?

The Divine Universe

In relativistic quantum field theory (QFT), matter and energy are no longer defined separately; both are included in the term, quanta. A quantum may represent matter or energy, depending on whether it is a "fermion" or a "boson". Fermions are the building blocks of matter, and bosons are the force-carrying particles which transfer energy. The quantum theorists have quantized, not only matter, but the forces—electromagnetic, strong, weak, and gravitational—as well; so that all energy interactions are constituted of bosons: photons that mediate the electromagnetic force, gluons that mediate the strong force, weakons (vector bosons) that mediate the weak force, and (the not-yet found) gravitons that mediate the gravitational force. Everything that exists is comprised of quanta, and is included in the Quantum Field. This Field is identical with what I have called "Divine Energy".

28.

APPENDIX 2: WHAT IS A SWAMI?

It's a question that comes up from time to time, and I've learned that I cannot really say what being a Swami means for all Swamis, but I can at least try to say what it means to me. I was living in a secluded cabin in the Santa Cruz mountains when it first dawned on me that I wanted to be a Swami. I had gone to live in that cabin in my spiritual quest for enlightenment, and I had been reading many books on Indian philosophy as well as books on Western religious philosophy. I was impressed by what Sri Ramakrishna's disciple, Swami Vivekananda said about *sannyasa*, and also by these words of Sarvepali Radhakrishnan: "A sannyasin [monk, swami] renounces all possessions, distinctions of caste, and practices of religion. As he has perfected himself, he is able to give his soul the largest scope, throw all his powers into the free movement of the world and compel its transfiguration. He does not merely formulate the conception of high living but lives it, adhering to the famous rule, 'The world is my country; to do good my religion'. Regarding all with an equal eye he must be friendly to all living beings. And being devoted, he must not injure any living creature, human or animal, either in act, word, or thought, and renounce all attachments. A freedom and fearlessness of spirit, an immensity of courage, which no defeat or obstacle can touch, a faith in the power that works in the universe, a love that lavishes itself without demand of return and makes life a free servitude to the universal spirit, are the signs of the perfected man." [1]

Well, who wouldn't want to be such a person? It was during this same period of time that I was given to experience a profound illumination from God, revealing the spiritual depth of my true being; and shortly thereafter, I made myself and God a promise: that I would first give myself a twelve year period of spiritual study and growth, then I would become a Swami. That was in 1966, and in 1978 I was able to fulfill that promise. After a paradisical five years in my cabin in the woods, I traveled to Ganeshpuri, India and became a disciple of the famous Kundalini master, Swami Muktananda.

Now, Muktananda (affectionately known by his disciples as "Baba") is known by many today as a man who made a tragic mistake in his later years by inappropriately sharing his physical affections with a number of young female disciples. Many of us will also make great mistakes in our lives,

The Divine Universe

especially as we age; and it is a terrible shame that Muktananda's great legacy of loving wisdom should be so tarnished by the memory of a few misdeeds in the latter period of his life. I was one of those who left his organization in protest and who spoke out condemning those misdeeds, and they needed to be condemned. But, because of those unfortunate events, few of the public today know of the greatness that was Swami Muktananda. His was a spiritual presence that touched the lives of hundreds, even thousands, of souls and lifted them to an experience of God in their lives through the generous gift of his own heart's immense compassion and love. Those who sat in his presence know as no others can that, despite his human imperfections, he was indeed a great saint, possessing immense compassion and awesome power.

In 1978, I was working in Muktananda's Oakland ashram, when I wrote to Baba in India informing him that the 12 years of my apprenticeship had expired and that it was time for me to become a Swami. He then invited me to Ganeshpuri to take part in the *sannyasin* initiations that were to take place in May at the time of his birthday. There were about a dozen of us, both Indians and Westerners to be initiated, and an appointed *Mahamandeleshvar* (ceremonial official) named Swami Brahmananda Sarasvati of the Shringeri Math was on hand to direct the proceedings. After performing the Vedic rituals of offering rice balls to our ancestors, and having the last remaining 'brahmin's tuft of hair' shorn from our heads, signifying the transcendence of all castes, we performed the culminating ceremony of discarding our old clothes while standing waist deep in a cold raging river at midnight, and the receiving of the Swami's ochre robes. After that, we were Swamis, monks of the prestigious Sarasvati Order.

But of course it is not the ritual ceremony that makes a Swami; it is the heart's desire, the commitment to a spiritually dedicated life, and the favor of one's Guru. I was to know the awesome power of Muktananda's grace to his Swamis, a grace that enlivened the world and my soul with a brightness that revealed God's sparkling beauty within and without. Through no merit of my own, I experienced a divine blue light that would indicate to me advanced souls by dancing over their heads; I would experience Muktananda's grace being emitted from my own body to sincere devotees; I was even able to experience the transference of spiritual energy to others when someone inadvertently brushed my clothes. It was all his amazing and gracious power, transmitted from him through me, even though he was not present. His loving regard of me, even from far away, was a tangible energy that drew me in awed devotion to know him as the very image of God and distributor of God's grace on this earth.

In Muktananda's organization, SYDA Yoga, Swamis were honored, not so much for their holiness, but for their position in the hierarchy of the

Guru's favor. Muktananda, in the tradition of the rajas of India, ruled as king over an orange silk-robed aristocracy or nobility, who always sat in the front nearest the king when he gave audience. Further back were the members of the functional bureaucracy, and behind them the peasants, the visiting mob. The Swamis shared in the teaching role, giving authorized courses and operating the regional Meditation Centers and Ashrams. In the absence of the Guru, they were the connection with the Guru and his teachings. In a way very similar to the monks and priests of the Catholic Church, the Swamis of SYDA Yoga made up an organizational hierarchy of representatives of the Siddha line.

But just as in the Catholic Church there were, and still are to some degree, lone contemplative hermits and anchorites who live among the people, in India there are many *sannyasins* who wander freely and independently, living the worshipful and contemplative life or teaching and lecturing and living by the charity of the citizenry. One can easily see, however, that such a class of religious itinerant beggars would not be feasible in Western countries. What, then, is a Western Swami to do? How is he (or she) to carry on his or her chosen vocation?

We must understand at the outset that a Swami transcends not only all caste designations, but all sectarian religious designations as well. A Swami is not (necessarily) a Hindu. The ideal Swami is learned in *all* religious traditions, and he is familiar as well with current science and literature. He is an enlightened and learned soul, and is solely dedicated to God and the well being of all God's children. After I had left Muktananda's organization, I was faced with the question of how to continue my "mission" as a Swami. My immediate instinct was to share my acquired experience and understanding in the form of writing, and I went on to produce a number of books, all concerned with the "mystical experience" and the Self-knowledge obtained thereby.

There was also, of course, the necessity of meeting the expenses of living in this world; and this I managed to do by obtaining a licence as a CNA (Certified Nursing Assistant) and working primarily as a Home Health Aide for elderly and infirm patients in their homes. For the twenty-five years since I left Siddha Yoga, I have written my books, seen to their publication, and daily served many of the victims of stroke, cancer, diabetes, kidney disease, and senile dementia with hands-on care. I no longer parade about in orange silk robes; rather, I live a simple solitary life as a servant; I promote my books, and I spend a good deal of time in reflection and inward communion with God. According to our brother, Socrates:

The Divine Universe

      This is that life above all others which man should live, holding converse with the true Beauty, simple and divine. In that communion only beholding Beauty with the eye of the mind, he will be enabled to bring forth, not images of beauty, but the Reality [Itself]; and bringing forth and nourishing true virtue, to become the friend of God and be immortal if mortal man may. Would that be an ignoble life?

—Plato, *Symposium*

        ஓ   ஓ   ஓ

A SONG OF THANKSGIVING

Hari, my love, I wish to sing to Thee a song of Thanksgiving.
Yet, O how I dread the futile search for meaningful words to offer Thee!
My heart is full of thanks and praise for each breath that is granted me,
But to speak reveals the lie of pretended two-ness that I must tell.
For Thou art my breath, my voice, the Real; and I am but the image;
I live by Thy uncommon Life, imaged in Thy dream of me.
And yet my gratitude to Thee upwells, as an image in a mirror
Might admire its own source, its real and original Face;
Or as a dream character might call out praise to its dreaming Self.

Though we are one, not two, I'll speak as though we're separate and apart;
For how else might I truly speak to Thee?
O Hari, Thou art alone, undiminished by the clatter and glitter
Of a billion billion images, mere reflections in a house of mirrors;
For Thou art alike the house, the mirrors, and the flitting images as well.
This speaking too is like the barking of a dog in an empty field;
For, though it may be heard, the silence of the cosmos remains unbroken.
Yet I, this imagined form, am present—at least in appearance;
And because I'm here, please let me speak to Thee in loving thanks.

O Hari, look how wonderful is this story Thou dost tell!
Look how beautiful is this body and the life ensouled.
Though all too quickly it will turn to dust, this form is Thine
And holds Thy greatness and Thy holy light and breath of life.
Thou, this brightly glowing wakeful knowing;
Thou, this deep and endlessly creative song of light and love
That bubbles up from Thy unfathomable depths
Within the soul of me to greet each day with joyful thanks.

O Hari, from Thy eternal Goodness and unknowable Repose,
Thou hast issued forth this universe of man and beast
With purpose known only to Thy own delight;

And Thou hast given Thy own thoughts to guide us from within
Through adventures great and small to bring us
Happily to our end in Thy boundlessly blissful Self.
O Hari, it is a most wonderful and admirable drama
Thou hast produced, full of harrowing dilemmas,
Frightful predicaments, and uproarious denouements!

Yet, in the end, we all awake to know one Self,
The Dreamer of this dream, our ever undisturbed Reality.
Always unperturbed, Thou art forever untouched by time,
As the patient sky is ever untouched by passing clouds;
We are where we have always been in truth, never separated
From our constantly enfolding, ever undivided Self;
Where all the fervent lives o'erpassed, like dreams,
Once left behind in waking, hastily retreat from view,
Revealed as the flimsiest of transient illusions.

In waking, we are one in Thee, O Hari!
And in Thee, as Thee, we have always been.
Never imprisoned as we thought in separate forms,
Once reawakened from our dreams, we know our
Ever undivided and eternal Identity as Thee.
In blissful folds of snow-white radiant Eternity
We rest as Thee in peaceful oneness and joy;
But while I live in pretended separation from Thyself,
Let me now offer my song of grateful thanks to Thee,
Who art the Life that lives me, my secret pride and joy;
For it is Thou who hast made Thyself as me.

Dear Father, all that Thou hast made is good,
And all Thy beauteous forms sing praise and thanks to Thee.
Then, let me uplift my voice in song as well
To glorify in praise my gracious Lord:
O Hari, all praise be to Thee in Thy heavenly glory!
All praise be to Thee in Thy universal pageantry of form!
My head is bowed in loving thanks and worship,
Knowing Thou art all and more than all.
Thy grace to me is beyond what my voice can tell;
I can but offer thanks, with hands held high, to Thee,
My ever kind and gracious Lord.

About the Author

Swami Abhayananda was born Stan Trout in Indianapolis, Indiana on August 14, 1938. After service in the Navy, he settled in northern California, where he pursued his studies in philosophy and literature. In June of 1966, he became acquainted with the philosophy of mysticism, and experienced a strong desire to realize God. Abandoning all other pursuits, he retired to a solitary life in a secluded cabin in the mountain forests near Santa Cruz, California; and, on November 18, of that same year, became enlightened by the grace of God.

He spent four more years in his isolated cabin, and subsequently met Swami Muktananda who visited Santa Cruz in 1970. Shortly thereafter, he joined Muktananda in India, as his disciple, and later lived and worked in Muktananda's Oakland, California ashram. In May of 1978, he returned to India and was initiated by his master into the ancient Order of *sannyas*, and given the monastic name, *Swami Abhayananda* (ub-hi´-uh-non-duh), "the bliss of fearlessness."

As a Swami, he taught in various cities in the U.S., including New York, Philadelphia, Chicago, and Oklahoma City; but in 1981, unwilling to condone what he saw as abuses of power, Abhayananda left Muktananda's organization, and went into retreat once again, this time for seven years, in upstate New York. It was during this time that many of his books were written, and Atma Books was founded to publish them.

At present, Swami Abhayananda is residing on the Treasure Coast of Florida, where he continues to work, teach, write, and publish his works on the knowledge of the Self. *The Divine Universe* is his ninth book. You may find all of his books listed on his website at: www.swami-abhayananda.com.

Printed in the United Kingdom by
Lightning Source UK Ltd., Milton Keynes
139375UK00001B/173/P